QB   Burgess, Eric.
581
.B87   Outpost on Apollo's
1993   moon.

$34.95

DATE			

BAKER & TAYLOR BOOKS

OUTPOST ON APOLLO'S MOON

OUTPOST ON APOLLO'S MOON

ERIC BURGESS

COLUMBIA UNIVERSITY PRESS   NEW YORK

Frontispiece: Opportunity beckons at the nearest new world.

Columbia University Press
New York   Oxford
Copyright © 1993 Columbia University Press
All rights reserved

Library of Congress Cataloging-in-Publication Data
Burgess, Eric.
Outpost on Apollo's moon / Eric Burgess.
p.   cm.
Includes bibliographical references and index.
ISBN 0-231-07666-5
1. Moon—Exploration.   2. Outer space—Exploration.
3. Project Apollo (U.S.)   I. Title.
QB581.B87   1993   919.9'104—dc20   92-26662
CIP

Casebound editions of Columbia University Press books are Smyth-
sewn and printed on permanent and durable acid-free paper.

Printed in the United States of America
c 10 9 8 7 6 5 4 3 2 1

Dedicated to:

Amanda, Brian, Cecily, and Chad
and all other grandchildren
of the space age.

Grasp the light
that rules the night
and with it write
the future.

CONTENTS

PREFACE

This book is not intended to be one on astronomy or on the geology of the Moon. It is about human expansion into space, an event that has been likened in importance to that prehistoric event when life emerged from the oceans and became adapted to living on the land. I want to emphasize the important role already played by our assault on the Moon. Also, how this should lead naturally to outposts there that will be equally as important to future human expansion into the Solar System.

Nevertheless I felt obligated to provide some background information about the Moon and its geology and about the early missions to our neighbor world. One cannot talk about post-Apollo missions without giving some background on what Apollo accomplished. In debating this point as I planned this book I was reminded of two incidents that taught me not to take for granted what is known by the recipients of my thoughts. In one instance, just before World War II, I had given a talk to a large audience about the possibilities of interplanetary flight. I was somewhat chagrined during the question and answer period when a young man asked quite seriously if I would tell him what was a planet. His question was followed by guffaws from the learned members of the audience, but it was quite valid. I had not explained what the planets were. Another instance was the famous, or infamous, depending on your viewpoint, case of *Rex versus the Manchester Interplanetary Society* in 1937. I had been summoned to account before a British magistrate for some infractions of the Explosives Act of 1875, and an Order in Council of Queen Victoria, associated with building and firing small experimental rockets. When the defense counsel referred to the Society at the beginning of the hearing, the learned magistrate asked, again quite seriously and somewhat incredulously, "Inter . . . what society?" So I ask my learned readers to excuse the rehashing of things they already know; to think instead of that half of the human population, the children of the space age, born after the first humans landed on the Moon. This rising generation will want to know what it was all about without having to refer to obscure scientific tomes if they are to vote for or against the base on the Moon. I hope that what I have to say will help them decide how they should cast their vote when it comes to matters affecting the future of spaceflight.

I see an expansion into the Solar System providing humanity with a vast new economic commons at a time when those of Earth are being lost to governments and international consortiums while a burgeoning world population is harassed by conservation, expensive energy, limits to growth, and fears of imaginary or exaggerated hazards from technological progress.

Since I have been personally associated with space activities and have crusaded since the 1930s for human expansion into space, I admit to being somewhat opinionated about many matters discussed in the book. My views may differ markedly from those of colleagues, of ecologists, groups at NASA, and many members of the public. I make no excuses for such differences of opinion since I am expressing firm personal beliefs—and that is, after all, the prerogative of a writer.

Space activities have been dogged from the beginning by a conflict between use of machines and use of people. Lunar exploration has suffered from conflicting opinions on how it should be accomplished. There was intense rivalry between unmanned and manned advocates, and between pure science and exploration. There is a need for both in each case. Unmanned exploration is vital to establishing details of the space environment and that of other worlds, and in making preliminary surveys of physical characteristics of those worlds. But we cannot leave the whole task of exploration to machines alone, unless we want to abdicate in favor of Von Neumann machines that will explore the galaxy, repairing and reproducing themselves in a bizarre imitation of a biological human system. I do not think we want to do that.

Currently we hear the old criticisms from reputable scientists that establishing a space station and bases on the Moon and Mars will stultify "real" science, wasting all available funds during these tough economic times. This is utter nonsense for the richest nation in the world, probably the richest nation there has ever been. When we see countless billions of dollars wasted in dubious commercial enterprises through questionable financial deals and "creative" accounting, when we see the cost of governing a nation increase orders of magnitude over the increase in population, when we see in every walk of life enormous sums wasted on advertising products which people do not need for a fulfilling life-style, the investment in a space station, a base on the Moon, and a base on Mars seems minimal by comparison.

I have pointed out in this book that the space station and the base on the Moon are by no means new ideas. They have been around for at least half a century, and both could have been practical accomplishments for the U.S. over twenty years ago. But having opened the extraterrestrial treasure chest and seen the wonders of its contents we slammed the lid tightly shut again. We are now far behind the Russians in space station technology despite their economic difficulties resulting from an unnecessary arms race, and we are losing our lead as a space-faring nation. This cannot be blamed on NASA for that organization has produced many plans over the years only to have them aborted because of lack of support either from Congress, the Administration, or the general public. And several in-depth studies of the nation's future in space have pointed the way quite clearly. Yet even today the space station program comes under repeated attack by columnists and members of congress. Consequently I have devoted some of this book to discussion of space stations and why they are important to future space activities, especially in connection with establishing an outpost on the Moon.

I have built much of my discussion about the lunar outpost and base on reports generated by various groups at NASA. However, it should be understood that these study results are not universally accepted. It has been my experience over many years of working with and writing about space and other advanced technology projects that human predictions must be subjected to a great deal of skepticism. We have all experienced this in reading annual reports of corporations, hearing promises of politicians, and being deluged by advertisements extolling the wonders that use of their products can supply. The same is somewhat true in looking into the future of technology and what might be achieved. A classic example is in H. G. Wells's novel *The War in the Air* (1908). Written soon after the Wright Brothers and others had built and flown heavier-than-air aircraft, Wells described future aerial conflicts. How ludicrous it now appears that he should accept future airplanes as continuing to be structures of stringers and wires.

I have seen that a pattern exists that I believe will continue; things get simpler going from the initial proposal to the actual doing. When an idea first surfaces in human consciousness, other people at that moment try to make executing it sound impossibly complex. But the doers eventually do it pretty simply! This may be because the original idea is the product of one or two minds working alone and butting against contemporary beliefs, while the doing is supported by the onslaught of many minds in what might be referred to as a group mind freed from restricting beliefs.

Nevertheless we must accept, as I certainly do, that no one knows what the future will be. Our oracles often turn out to be false prophets swinging wildly between extremes of doom or bonanza. We can only speculate about the future and accept that we may be hopelessly wrong. But unless we speculate, as I have done in this book and other books and in many magazine articles for half a century, and speculate in a form that will develop public interest to urge politicians to support human expansion into space, we shall never get a base on the Moon. As far as the American people are concerned, without continued expansion into space we will be sentencing our children to be stillborn in the womb of Earth while other nations venture forth into the great outside and strive for adulthood of our species.

Make no mistake, the rigors and cost of human expansion beyond Earth are enormously challenging, but equally as challenging proportionally were the rigors and costs of human expansion from the European continent. And today the nations of Earth have enormous wealth much of which is squandered on amassing unprecedented military arsenals to defend themselves against themselves.

Although opinions expressed in the book are my own, I must acknowledge my great appreciation of the contributions I have derived over many years from people I have interviewed; scientists, engineers, and members of the general public. I am also indebted to public relations or public affairs officers of many companies, universities, and NASA facilities. Many of the illustrations came from aerospace companies during the Apollo and other lunar programs. For these I am extremely grateful and they have been acknowledged individually where they are reproduced. One unforgettable aspect of the lunar program was the readiness of the many companies involved to make available for publication the details of their contributions to what was certainly a most momentous episode in human history. Since many thousands of companies were involved I must apologize to those I could not mention. But Apollo was undoubtedly a triumph of U.S. private enterprise; for examples, small machine shops that developed extremely innovative ways to make a multitude of new parts, textile workers and seamstresses who made out-of-this-world suits to protect humans in the vacuum of space and on the Moon, great aerospace companies that placed their best engineers and scientists on developing ultra-reliable machines of unprecedented size, universities that encouraged a new breed of students pushing at many unexplored frontiers of knowledge, and medical researchers who found ways to safeguard humans in an utterly alien environment.

And most important, supporting this unprecedented activity was the great mass of thoughtful people who thrilled to the vision of a human race no longer confined to the restrictions of planet Earth. The preliminary human exploration of the Moon was the greatest effort, short of war, that a nation had ever undertaken. It was truly a realization of an impossible dream, a veritable miracle of technology and organization directed toward a common and clearly defined national goal.

Now we have the opportunity to revive that spirit of adventure by accepting and implementing the new national goals in space. Now is the time to excite our youth with vistas of new explorations, stimulate our economy with worthwhile activity that will be a solid investment in the future of our nation, our species and, in the long haul, possibly of terrestrial life forms in general. For the Earth is not a stable planet; its continents and oceans are constantly changing. Beneath a fragile crust a seething ocean of hot magma seeks to break loose and spread across ocean floors and continental masses. While humans squabble interminably over tracts of land and boundaries invisible from space, Gaea laughs at our futile efforts at domination, perhaps the sole survivor of the ancient Greek pantheon.

Now little gods on modern Mount Olympuses hold their summits with nuclear-tipped thunderbolts of Zeus poised deep within missile silos. We mortals await their pronouncements as though those few representatives of the new pantheon have absolute control over our destinies. The mass media marvel while individuals of the current pantheon maneuver for power among themselves as ardently as did those lesser gods of the old. But as the old pantheon of that ancient Olympus faded into obscurity, perhaps we see already the beginning of the demise of the most recent pantheon that human thought has deified. The collapse of the mutant communist empire and the tottering worldwide of errant capitalists in a plethora of financial scandals may signify a new beginning, an end to summits and a return to meetings of unified common people.

Perhaps a united international effort to establish transportation nodes in low Earth orbit, as described in this book, followed by outposts and bases on the Moon, will be the stimulus for a new period of human activity. Can we hope for an era of mutual assistance on a global scale, including a modern equivalent of the industrial revolution, but beyond Earth so that our delicate ecology is preserved and the evils of the old industrial revolution are avoided? Perhaps we can hope for and achieve a new expanse of broadened horizons in which many more humans of all walks of life will have an opportunity to stand on other worlds and observe the material universe afresh. Perhaps as a species we will begin to obtain a much broader spiritual consciousness of reality and of all other humans as being of one family. Maybe we shall practice cooperation in place of domination, accept unity in place of division, and acknowledge common instead of separate roots.

But even without such idealistic results, which I must acknowledge may be utopian, the economic benefits of a switch from international arms marketing to a concerted effort of exploration beyond Earth will be well worth the investment.

At least, that is how it seems to me. Whether you agree or not, I hope you will enjoy thinking about my speculations on how developments of our space technology might lead your children or grandchildren to stimulating adventures and experiences in some kind of lunar city. It may make up for leaving them with an astronomical national debt. I sincerely hope so.

Eric Burgess
Sebastopol, California

OUTPOST ON APOLLO'S MOON

1 EXPLORERS: THE INCURABLE OPTIMISTS

On the occasion of the first landing of American astronauts on the Moon in 1969, a night some 25 years earlier came vividly to mind. I was with about a score of other space enthusiasts meeting in wartime London to plan how we could urge humanity to send a mission to the Moon after hostilities ceased. We sat at a long table in an upstairs room of the Northampton Arms, a pub in Islington a few miles north of St. Paul's. There was an awful racket from the public saloon downstairs. Refugees from the outer darkness of the ubiquitous blackout drowned their wartime sufferings in warm British ale and lustily sang wartime morale boosters.

As a frightening crashing rumble blasted over the words of "Roll Out the Barrel," we looked knowingly at each other. The awesome noise of an exploding V-2 rocket warhead nearby brought a gleam of anticipation rather than a cloud of terror to our eyes. It was our proof that the job we had so long dreamed of could be done. Large rocket vehicles could be made with unprecedented lifting power. Liquid oxygen and fuel could be tamed in big liquid propellant rocket engines to thrust humankind into space. Our dreams of flights to other worlds could indeed become realities.

About 11 years before this episode at Islington the British Interplanetary Society had been formed by Phil Cleator in Liverpool, a major British seaport. We of the interplanetary movement had maintained a loose organization during the war orchestrated by two other astronautical groups that had banded together into a group known as the Combined British Astronautical Societies to preserve some continuity. We met whenever we could at semi-official meetings in Manchester, Birmingham, Farnborough, and London, and often individually. I remember Arthur C. Clarke and I spent a very rainy afternoon wandering around the grounds of historic Warwick Castle, one of Britain's finest Medieval castles, talking about how we could get a strong interplanetary society going and use World War II technology to send humans to the Moon. And I remember visiting Phil Cleator in Wallasey soon after his parents' home had been demolished by a Nazi bomb with the loss of his mother and father and all his records and papers. Phil and I discussed how the interplanetary movement might be organized after the war to push humanity out into space in a new program of far flung exploration of the Solar System as suggested in his pioneering book (*Rockets Through Space,* George Allen and Unwin Ltd., London, 1936).

When we interplanetary enthusiasts were able to get together as a group in London we found the only meeting rooms that were readily available were those in pubs. In those somewhat dismal upstairs rooms we dreamed of the day when we could meet at scientific institutions, when people would take us seriously, and when flights into space and to the Moon would not be shot down before they blasted off by those who should have known better; the professional scientists and engineers of repute, notably astronomers and aeronautical engineers.

Our interplanetary group was quite a select cadre of individuals of entirely different backgrounds (figure 1.1). By the early 1940s it included Val Cleaver of de Havilland who would later, at Rolls Royce, lead the development of rocket engines to power the ill-fated British Black Knight booster. There was John Humphries, then working at Farnborough's Royal Aircraft Establishment and founder of a small interplanetary group there, who later went to the British Atomic Energy Commission. Arthur C. Clarke and I were both in the Royal Air Force; he was in radar research while I was an instructor in radio communications. We had individually scooped first and second prizes respectively in a Royal Air Force Quarterly essay competition about the role in future warfare of guided weapons and ballistic V-2 type missiles.

Kenneth Gatland was with Hawker Aircraft and later made many contributions to the literature of spaceflight. Arthur Janser was a chemist who had first proposed the use of solid propellants in a cellular step arrangement to achieve high mass ratios for the Moon rocket. Ralph Smith and Harry Ross had already produced many innovative technical ideas needed for a moonship and later produced designs for soft-landing vehicles and for lunar spacesuits and Earth-orbiting manned space stations. Professor A. M. Low, a somewhat controversial figure, had long conversations with me about the future of the interplanetary movement in Britain. Through his prewar magazine, *Armchair Science,* he had done much to help the society along in its early days.

We were quite the incurable optimists. Many of our contemporaries would unkindly relate us to a lunatic fringe, way out dreamers who would never live to see their dreams become realities.

For example, Professor J. B. S. Haldane once wrote to the Society just before the outbreak of war stating "While at the moment I think it more important to concentrate on saving our existing civilization than on journeying to the moon, I think it likely that your efforts and those of similar organizations elsewhere will pave the way to success within the next thousand years."

FIGURE 1.1 Prominent members in the British enthusiasts urging for a voyage to the Moon in the 1930s. Members meet at the inaugural meeting of the London branch of the Society on October 27, 1936, at A. M. Low's office, Waterloo Place, Piccadilly. The members are: back row, A. Janser, third from left; Ralph A. Smith, sixth from left. Far left on the front row is Miss E. Huggett, the secretary of the British Interplanetary Society for many years. Next to her is J. H. Edwards, research director. In the center is Professor A. M. Low; next to him is W. H. Gillings; then Arthur C. Clarke, sixth from the left; next to Clarke is E. J. Carnell who chaired the meeting. (*Photomontage in Burgess archives from original photos by L. Klemantaski*)

The prestigious science magazine *Nature* was not so kind. Its editor did not even guess at a thousand years! Commenting on the 1939 issue of our Journal, which contained detailed ideas for a moonship, the editor thought that the whole affair should be dismissed as a "wildcat" speculation, claiming that "little confidence could be placed on the deliberations of the British Interplanetary Society."

An even more damning criticism came as late as 1941 in a mathematical paper published in the *Philosophical Magazine* in January of that year. The author, a president of the Royal Astronomical Society of Canada went to great pains to "prove" mathematically that to send a spaceship to the Moon and back would require a vehicle weighing a trillion tons at takeoff. Clearly this was an engineering impossibility. But all the calculations were based on obsolete information used by the author. He ignored the work done over the previous two decades and assumed that the mass of the spacecraft returning to Earth would have to be 500 tons and the rocket engines for the task would develop a jet velocity of only one mile per second. As Apollo demonstrated later, a successful Moon voyage could be accomplished with a returning spacecraft weighing a few tons (similar to that proposed in the 1939 British Interplanetary Society study), and even the V-2 was by then developing a jet velocity of 1.27 miles per second with a liquid oxygen/alcohol rocket engine. Use of liquid oxygen/liquid hydrogen propellants pushes the jet velocity higher with impressive reductions to the initial launch mass. By the time of Apollo the F-1 engine for the Saturn V launch vehicle developed a jet velocity of

1.6 miles per second (2.57 kps) and the J-2 engine one of 2.5 miles per second (4.02 kps), while a proposed engine for a new heavy lift launch vehicle would have a jet velocity of 2.75 miles per second (4.43 kps). But the critics of spaceflight were as vociferous in their condemnations of ideas of traveling to the Moon as were earlier critics ridiculing heavier than air flight before Kitty Hawk.

After years of war, we of the British interplanetary movement were feeling very much alone at the time of the initial V-2 flights. There were rumors in the British intelligence community that the Nazis had developed a nuclear weapon to be used as a warhead for the V-2. The Hitler regime had scattered the Germans we used to know; some had fled to America, others were scooped into the top secret German war machine. Even our friends in the U.S. appeared to have developed cold feet about interplanetary flight and were concentrating on rockets and jet propulsion for more mundane terrestrial applications.

But, as Cleaver had written in 1939: "Those who believe that space flight is still centuries off might ponder the fact that only 60 years separated Henson and Stringfellow [designers of early models of heavier-than-air aircraft] from the Wright aeroplane." Indeed the Royal Aeronautical Society was formed in London 50 years before the Wright Brothers flew their craft. We believed that man would fly to the Moon in our lifetimes, and the belief was justified when Neil Armstrong stepped onto the surface of the Moon only 35 years after the formation of the British Interplanetary Society. Our dreams in that dimly lit room of wartime London had been realized. It seems we knew deep down in our consciousness that what may be regarded as impossible one day may be commonplace soon afterward and that real progress is made by dedicated people who believe and not by detractors who try to impede all new ideas. I recollect that one of our members, most probably Arthur C. Clarke, commented that our descendants will not laud us for traveling into space and landing people on the Moon, but will criticize us for having taken so long to do so.

In 1969 I was science correspondent for an international newspaper; I interviewed the geologist Eugene Shoemaker who was expressing his dissatisfaction with the way the Apollo program had been conceived and implemented. He stated that: "Exploring is extending man's knowledge of his surroundings . . . and extending his capabilities." He contended that the space program should be analogous to the systematic scientific exploration of the Western states during the last century, such as the King and Powell surveys. It should not be a political expedient to develop a transportation system to demonstrate technical superiority over other nations. The King and Powell surveys were planned very carefully to be efficient scientific expeditions, to explore the country and to map the topography.

Said Shoemaker, "I think that we can show that it is worth sending a man into space when we really know how to use him effectively. The real goal of the United States in space is to explore . . . and specifically the main goal is to explore the Solar System."

"It is exploration that makes man different from the rest of the animal world," he continued. "And it [exploration] has led man to what he is."

"If you look back in history it is those societies and those nations which have explored which have been the vigorous societies."

His philosophy applies even more so today. The role of NASA should not be

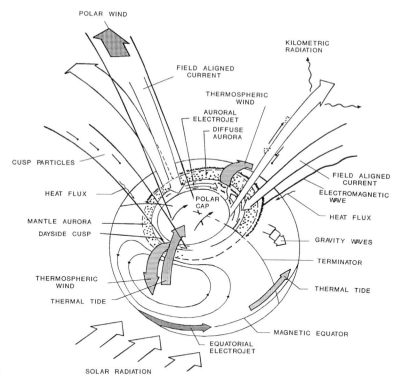

FIGURE 1.2 The dawning of the space age threw an entirely new light on planet Earth. Spacecraft revealed the complex nature of the environment surrounding the planet; an intricate pattern of electromagnetic fields interacting with a blizzard of protons and electrons streaming through space from the Sun. Much of the environment within the atmosphere, biosphere, and hydrosphere may be governed by this dynamic region of particles and fields revealed by the space explorers. (*NASA*)

solely to develop new transportation systems, as was the case with Saturn/Apollo and the Space Shuttle, but it should be to organize for a long-term and sustained program to explore the Solar System and investigate it scientifically. Then following the initial scientific exploration, the Solar System will offer opportunities for free enterprise to develop its resources in much the same way that traders brought about a global economy following the initial military and scientific explorations on the Earth. As with the shipwrights of that earlier age of exploration, engineers to develop space transportation systems are absolutely necessary, but the purpose of such systems must be clearly defined, just as aerospace companies develop airplanes to meet the needs defined by the airlines who use them in day to day operations. Transportation systems are never an end in themselves but are merely a means to the end of broadening human awareness of the total environment. This totality necessarily includes not only the environment of our own planet but also that of the Solar System and, looking farther afield, of the galaxy in which all terrestrial life forms, the Earth, and the Solar System reside.

Moreover, an appreciation of the intricacies of this total environment may be of much greater importance than the restricted environment of the planet itself (figure 1.2). Recent research has pointed toward dangers from beyond the Earth: violent solar disturbances, loss of protecting magnetic fields and atmospheric layers, and the onslaught of mountain-sized bodies that frequently pass very close to Earth at 40,000 miles per hour (64,000 kph) and, it is believed, every 100–200 million years or so crash into the planet and effectively eliminate all life forms with but a few exceptions. Protect the wildflowers if we will in vernal pools, or the owl in the ancient forest, but in the long term both they and we have a very good chance of being eliminated by chance catastrophes from space of which we now have very little basic knowledge. Only by exploration might we find answers and somehow develop means to revise our ways and wisely protect all terrestrial life forms from a cosmic catastrophe. Remain bound stillborn in the womb of the planet we will otherwise experience a completely unpredictable annihilation, unwarned and unprepared for the cosmic Armageddon.

Despite all the criticism, the quest to reach the Moon continued, stimulated by an early success of Soviet Russia in sending a spacecraft around the Moon and returning pictures of the far side to Earth (figure 1.3). In 1959 the U.S. Advanced Research Projects Agency (ARPA) authorized funding for several unmanned flights to the Moon. Three lunar missions were to be undertaken by the Air Force's Ballistic Missile Division using intermediate range ballistic missile (IRBM) rockets as launch vehicles equipped with several upper stages to send a small payload to the Moon. Two were to be the responsibility of the Army using the Jupiter C configuration that had launched the first American satellite. The lunar program was designated as part of the U.S. activities for the International Geophysical Year.

At the press room in Inglewood, California, location of the headquarters of the Ballistic Missile Division, I was among the crowds of journalists who were to be witnesses to the first U.S. reaching for the Moon. We were disappointed. In October 1958 the first lunar shot (figure 1.4) attained only just over 72,000 miles (115,870 km) from Earth. The second shot fared even worse, attaining a distance of only 963 miles (1550 km) in November of that year. The third shot came under the jurisdiction of the newly formed National Aeronautics and Space Administration (NASA), but in December 1958 it reached only 63,580 miles (102,320 km). Renamed Pioneer, the fourth lunar shot did escape Earth and passed within 37,300 miles (60,000 km) of the Moon in March 1959, and Pioneer 5, a more complicated spherical spacecraft, hurtled by the Moon to enter solar orbit in March 1960. The groundwork had been laid. Subsequently there would be Ranger and Surveyor and Lunar Orbiter, all leading toward the crowning achievement of Apollo and landing humans on the Moon and returning them safely to Earth.

It was an exciting era the likes of which may never be repeated, since it brought forth our species into the wider universe of extraterrestrial activities with all its immense challenges. The missions to the Moon brought a new view of Earth, while closeups of the material of the Moon changed many viewpoints about that challenging world and its history. Fred Hoyle of the University of Cambridge, a noted British astronomer, was principal speaker at a banquet hosted by the National Space Hall of Fame and the City of Houston for more than a thousand scientists gathered in January 1970 to discuss the results of the preliminary analysis of samples returned from the first human expedition to the Moon. Hoyle, while admitting that he had not been an early supporter of spaceflight and had doubted that humans would set foot on the Moon, nevertheless reminded his audience that he had

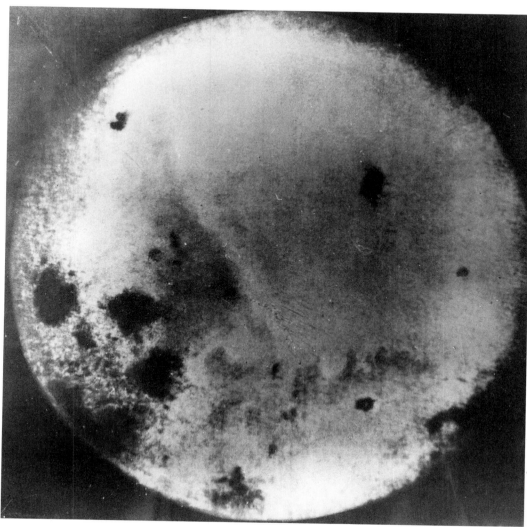

FIGURE 1.3 The first time humans saw the far side of the Moon was when the Soviet Lunik III spacecraft returned images in October 1959 to reveal a surface quite different in many ways from the Earth-facing side. (*TASS*)

predicted in 1948 that: "Once a photograph is obtained of the Earth from outside . . . a new idea as powerful as any in history will be let loose."

At that time I reported the story for a national newspaper, writing that Hoyle had asserted, "Quite suddenly everyone worldwide has become seriously interested in protecting the environment." He pointed out that previously such thoughts had been largely ignored. But he said that it was significant that in the year when men first set foot on the Moon the desire to save their own planet blossomed, a new perspective has been gained from manned spaceflight. "We are in a transition period," he added," and in the future we must rely more on outside laboratories than astronomers did in the past." He said that continued examination of lunar materials would tell us more about the Solar System than anything we can accomplish on Earth.

FIGURE 1.4 First American attempts to reach the Moon began in 1958 when the Air Force launched several spacecraft using modified ballistic missile rockets topped by upper stages. This spacecraft, Pioneer I, did not reach the Moon but gathered information about the extent of the radiation belts of Earth and measured impacts of micrometeoroids. (*U.S. Air Force BMD*).

Hoyle predicted that long after the problems of today are forgotten, our descendants will remember that humans had landed on the Moon in the 1960s.

Soon after the first Apollo mission I was privileged to inspect lunar materials through a microscope and I saw the intricate details of the previously undisturbed lunar surface revealed in 3–D images. There before me was a piece of lunar soil gathered from the Sea of Tranquility and set up under a microscope at Caltech. A several billion year old dark sphere of glassy substance was highlighted by a brilliant spot of light from the fiber optics illumination. It seemed strange to peer at the Moon through a microscope when most of my life I had only looked at it through telescopes of various apertures. There were other intriguing bits of the Moon to be inspected later; glassy streaks of material, glass–lined pits on fragments of rock, and microscopic splashes of glass adhering to other lunar particles (figure 1.5). All the particles appeared alien, quite unlike any bits of terrestrial soil, nevertheless composed of much the same basics: silicates and other rocky materials. All appeared to have been modified by intense heat, the cataclysmic mechanical forces of enormous impact, and the searing rain of solar and cosmic ray particles.

A short while later, at the Boeing Research Laboratories in Seattle, I had another look at the Moon revealed by the Apollo mission, a closeup view of the beauty and intricacies of a lunar surface preserved almost intact for eons of time; a jeweled pattern of delicate "fairy castle" structures unlike anything seen here on Earth. While others marveled at the impression made by Neil Armstrong's boot as he took his first step on the dusty surface of the Moon, I was enthralled by this lunar world of beauty that was generally hidden from human eyes. Each 3–D picture that I inspected through a stereo viewer showed a surface area that measured about three inches square. Within this small space I saw miniature peaks crowned with jeweled scintillations of gold and silver hue. Others were splashed with iridescent greens and yellows. There were also rocky areas marked with glassy splotches and some areas with deep cracks that had intriguing undisturbed depths penetrating between rocks like abyssal caverns. A world of intricate colored beauty rivaled the most colorful of terrestrial undersea coral gardens I had seen.

But these delicate spires, jeweled mountains, and mysterious caverns, molded by many millions of years onslaught of particles of the solar wind, were invisible to the eyes of the astronauts. Their beauty was not hinted at in the early panoramic photographs of the lunar surface nor in the rocks brought back to Earth for inspection. The astronauts obtained the images with a special stereo camera mounted on a probe. The camera recorded the surface undisturbed by humans, revealing how it had been molded by many millions of years of exposure to the vacuum and the radiation of space. It seemed a shame that inevitably the armored feet of the astronauts and the wheels of their vehicles would pulverize many of these delicately colored intricate sculptures into nondescript dark grey powder to be wafted over the Moon's surface by the rocket blast from the Lunar Module as it returned its human cargo of explorers to their home planet.

The Apollo program, expanding dramatically and often unexpectedly on earlier lunar exploration programs of Ranger, Surveyor, and Lunar Orbiter, and of the Soviet Luna and Lunokhod, provided access to an encyclopedic storehouse of information that has been likened to the gift of a planetary Rosetta Stone. As a result of this manned exploration we can now use this information to decipher many heiroglyphs of planetary evolution telling, at least in part, the story of the Earth on which we live and its relationship to the other planets and to the rest of the universe. Unmanned missions alone might never have gathered such a wealth of new information.

The Moon has remained almost unchanged throughout the period of planetary history recorded in the rocks of Earth, and before life developed here. The lunar records have already provided leads toward unraveling the mystery of origin of the terrestrial-type planets, the materials of which they are built, and the energy sources that drive them. But many questions remain, the answers to which depend upon a return to explore the Moon with long-term human presence on its surface to permit extended exploration. Many geologists suggest that the Moon, orbiting there so tantalizingly close to Earth, has many more surprises for us. In effect, scientists have only scanned some pages of the Moon's encyclopedic history; now they need intensive and penetrating study for real understanding. The surface of the Moon

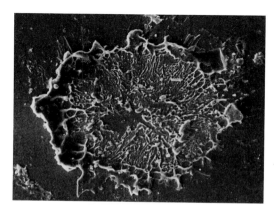

FIGURE 1.5 When viewed under a microscope, lunar rocks display violent impacts in which glasses spread across the surface, as shown in this electron micrograph of a splash of glass materials on a submicroscopic glass bead. (*NASA*)

also preserves a rudimentary cosmic record not only of the Moon itself but also of the Sun and possibly of the galaxy.

Before the space program blossomed into fulfillment we had only two samples of matter to analyze directly: the material of the Earth and the material of meteorites that fall to the Earth. The latter are believed to be fragments of material from within the Solar System, possibly remaining samples from the primitive solar nebula. A general class of meteorites called chondrites appeared to contain the chemical elements—except for the lighter volatile elements—that spectroscopic analysis had shown to be present in the atmosphere of the Sun and stars. Thus the chondrites could represent primitive material from which the Solar System formed.

Astronomers had calculated that the density of the Moon was less than that of the Earth, and they had assumed that the Moon contains fewer heavy elements than does the Earth. In fact, the density of the Moon was about that of the chondritic meteorites. Some scientists contended that the Moon consisted of chondritic material and would everywhere be the same since it seemed likely that it had formed cold from the infall of those bodies. Thus it was believed that the Moon would have the same chemical composition throughout its bulk and would effectively be a very uninteresting large chondritic meteorite on which there had been no volcanic activity; the flat mare areas might be dried lava beds derived from energy of impacts and later smoothed by water.

Still, other scientists argued that the Moon had heated following its accretion and any initially chondritic homogeneous mass would have separated into layers when light elements rose toward the surface and heavier elements sank toward the center. They argued that volcanic lavas had later flowed on the surface and filled the low-lying areas.

Before the lunar programs two explanations for the origin of the numerous craters seen on the surface of the Moon were hotly debated in scientific circles. One group contended that the craters were the result of impacts by projectiles falling onto the Moon from space; another contended that they were volcanic in origin like calderas on Earth.

The results of the missions show that most of the craters originated from impacts, though later some volcanic activity did take place. However, the argument is not settled to everyone's satisfaction; there are still a few scientists who contend that volcanic features on the Moon's surface are revealed by photographs from the Apollo Command Modules that orbited the Moon. Today it seems most likely that there was some volcanic activity, but the question remains how recently it ended.

While Apollo and other lunar missions showed that the surface of the Moon has been molded mainly by impacts (figure 1.6), they also confirmed that extensive lava flows (figure 1.7) had later filled the floors of craters and the big circular basins. The main lava flows appear to be volcanic, although it seems quite possible that a melting of rocks also occurred from the energy released by impacts. This is particularly evident, for example, in the inner walls of fresh impact craters such as Aristarchus.

That there has been heating and melting of the Moon now seems clear, but prior to Apollo many scientists had accepted a lunar theory of cold accretion from the primordial solar nebula in which the interior of the Moon never became hot enough to melt. However, the Apollo astronauts gathered data that have shown that the

FIGURE 1.6 A large proportion of the lunar surface is covered by overlapping craters, the evidence of an intense bombardment from space some 4 billion years ago. In this image of the far side, acquired from Lunar Orbiter, a rectangular-shaped crater, located to the right of the central peak of the large crater, is unusual. But to its right another larger crater has a straight wall also. These linear formations probably result from faulting of the lunar surface. (*NASA*)

Moon most probably melted soon after its formation and that a crust floated to the surface. The Moon also rapidly cooled as it continued to be bombarded by projectiles from space. Radioactive elements were concentrated within upper layers by the initial melting and differentiation. This led to a second stage of heating in the upper mantle and crust that, in turn, produced the lava flows so prominent on the Moon today. This second melting took place about one billion years after the first. From that time the Moon has been quiescent.

General debris is deposited over the surface of the Moon as a lunar regolith (figure 1.8), a mixture of pulverized soil and rocky fragments. This layer is churned by incoming meteorites in a "gardening" process. The lunar regolith varies in thickness. On the flat lunar plains it is a few meters thick. At the edges of rilles and on mountain tops and fresh crater rims it is much thinner. In some places the bedrock is exposed, for example at the edge of the Hadley Rille where bedrock was sampled by the Apollo 15 crew.

The regolith consists mainly of rock types churned up from the material beneath it. In addition it contains materials from remote locations that have been transported across the surface by major impacts, glass that has been produced by impacts, and the material of the infalling meteorites.

FIGURE 1.7 Prominent on the Earth-facing side of the Moon are the relatively flat areas of dark lava flows that flooded ancient basins and abut on the surrounding mountains. This Lunar Orbiter picture shows the region of southwestern Oceanus Procellarum with the crater Damoiseau and a sharply defined contact between the mare surface and the uplands. (*NASA*).

The layering of the regolith provides a historical record of the bombardment of the lunar surface by meteorites, solar particles, and cosmic rays. There is even some recent evidence that this regolith carries the record of the Solar System passing through the spiral arms of the Galaxy at three separate periods about 1 million years apart during the Precambrian era of Earth's history.

In 1950 scientists generally accepted that the age of the Earth and of the Solar System was about 3 billion years. This was based upon geological evidence from terrestrial rocks, the age of meteorites determined by radioactivity analysis, and theoretical analysis of the development of the Sun and its production of energy. Such an estimation was a great advance over the nineteenth-century physicists' arguments, notably those of Lord Kelvin who, before the development of radioactivity measurements to determine the age of rocks, had insisted that the Earth could not be more than 100 million years old. Even Kelvin's estimate was far longer than those of earlier philosophical and religious arguments that contended that the Earth was only a few thousand years old.

Scientific analysis of radioactive isotopes in rocks pushed Earth's age to at least 3.7 billion years just before the space program, but samples obtained from the

Moon by Apollo threw an entirely new light on the age of the Moon and consequently of the Earth and the Solar System.

The lunar samples revealed ages of more than 4 billion years, a time at which 90 percent of the impacting bodies that blasted out the lunar craters had fallen on the Moon. The extensive lava flows into the mare basins occurred between 3.2 and 3.8 billion years ago. By contrast the oldest known Earth rocks are generally about 3.7 billion years old, although some recent research by Ken Collerson of the University of California at Santa Cruz indicates that samples of mantle rocks from outcrops on the Labrador coast may be 4 billion years old according to neodymium/samarium ratios. The present ocean basins on Earth are very much younger, having been formed less than 300 million years ago. The lunar rocks also show that most of the

FIGURE 1.8   All the surface of the Moon is covered by a layer of shattered rocks and churned-up material of lunar soil referred to as the regolith. This photo obtained by the crew of Apollo 17 shows the surface in the region of Taurus-Littrow Valley. (*National Space Science Data Center*)

activity that molded the surface of the Moon to its present form occurred more than 3 billion years ago. Therefore, the Moon's surface provides a record of events in the Solar System predating most of the terrestrial rock records available for analysis.

Surprisingly large amounts of very ancient glass were discovered on the Moon. On Earth there are only a few natural glasses, such as obsidian, and these are not more than 200 million years old because glass vitrifies, i.e., looses its glassiness, when exposed to a water-rich environment. Lunar glasses are present in abundance with ages of billions of years. Their presence indicates that the Moon never had significant amounts of water.

Although no free water was expected on the Moon, the absence of water bound in lunar rocks surprised scientists. Many had anticipated that, as in Earth's rocks, it would be found in the lunar minerals. Virtually no such water was found in any lunar sample.

Today the Moon has an extremely weak magnetic field only. Its rocks, however, have an unexpected remanent magnetism, implying that the Moon was exposed to a large magnetic field 3 to 4 billion years ago. Whether this was the field of the Moon itself, similar to Earth's intrinsic field, or some external field through which the Moon passed as it was cooling, cannot be determined.

The Moon is quiet internally. Moonquakes are ten thousand times less in number than earthquakes; and moonquakes are very weak compared with earthquakes. But the amount of heat flowing from the interior of the Moon to its surface is abnormally high for what seems to be a relatively cold, quiet, and dormant body. The heat flow is half that of the terrestrial heat flow, which indicates that the Moon is vastly enriched in radioactive uranium compared with the Earth. Yet the lunar radioactive elements must be concentrated toward the surface; otherwise the interior of the Moon would be molten today and more seismic activity would be expected.

Exploration of the Moon has led to an understanding of much more than of the Moon itself; it now provides an important tool of comparative planetology by which scientists can compare the origins, evolution, and present states of the planets of the Solar System and why Earth is such a unique planet.

The detailed information obtained from scientific experiments on the surface of the Moon coupled with the direct observations of astronauts there have been systematically connected to information obtained by remote sensing from orbit about the Moon. Thus the lunar science experiments have shown that valuable geophysical and geochemical information can be obtained from orbit without an actual landing on a planetary surface. Although remote sensing from orbit had been tried before the Apollo experience, scientists were not sure how valid the results were. After Apollo they knew that these techniques can be applied to investigate other planetary bodies without necessarily having to land upon them. Already the techniques are being applied to survey Earth's resources from space. They will be important tools for the exploration of Mars, as they will help to select areas to be explored by automated roving vehicles that will be used to choose a site suitable for the landing of humans and establishing a base.

Like all major exploratory efforts, the Apollo program was just a beginning. The information and samples obtained will be used by generations of scientists just as in the past samples and data from explorations on Earth were used for many years after the actual explorations. But in the past the samples often became useless because of environmental pollution. Priceless Egyptian artifacts that had lasted centuries in the dry climate of the Nile Valley rotted in humid museums of Europe and America. Paintings exposed from sealed tombs deteriorated rapidly. Meteorites captured as fresh fragments from space absorbed terrestrial pollutants that masked their cosmic stories. So while the technology available to examine samples and wrest from them important information rapidly progressed, the samples themselves deteriorated and much of the new technology could not be applied.

Today we have a priceless collection of samples from the Moon. Although they seem to be tough rocks that will last for ever, a short exposure to the atmosphere of Earth would erase from them many of the finer details of their cosmic record, analogous to a high quality cassette recording being exposed to a magnetic field. Without providing a rigorously controlled environment to store the lunar samples we would be denying later generations of scientists the opportunity to use better techniques than we possess today to try to unravel the chronological, chemical, mineral, and petrological records.

The rocks are now available for study to qualified scientists under conditions that prevent their contamination. Some specimens have been made into thin sections to be inspected under a microscope. These sections are available for study at universities and, as colored slides, at high schools. Thus the results of the lunar sampling program are available to many practicing scientists and students.

The analysis of the lunar samples to date has also had an important impact on terrestrial geology and geophysics. New age-dating methods developed to investigate the lunar rocks are being applied to look into the history of terrestrial rocks. The Apollo science program fostered a new type of international cooperation among analytical laboratories. Samples were sent to laboratories worldwide and yielded reconcilable results—something that was almost unheard of prior to the lunar science program.

In the decade of Apollo exploration an entire new world was suddenly opened for study and found in some ways to be similar to Earth but in others entirely different. There was probably more excitement among many hundreds of international scientists attending the lunar science conferences in Houston shortly after the return of astronauts from the Moon than at any other scientific conference during the present century.

These large gatherings expressed the excitement of exploration of new and challenging frontiers that is both a stimulus to the individual mind and an essential food for the spirit of the whole of humankind. The lesson of history has been that without exploration humans atrophy, nations degenerate, and whole civilizations collapse.

The missions to the Moon of the National Aeronautics and Space Administration revolutionized human thought about this neighbor world and rendered nearly all previous speculative theories obsolete. The consequences were, however, farther reaching than lunar science; they applied to all science. The surprises from Apollo reinforced the truth that science is not theory or speculation, no matter how thought provoking such theory or intriguing such speculation may be. Science is interpretation of observations derived from exploration. It is always only a "best for the time being" explanation of what we observe; an unending challenge of applying impatient curiosity to understand our environment ranging from the fundamentals of matter and radiation to the congregations of galaxies. We try to find evidence of natural laws, of reason, and of order in which we can participate. We try to establish rules of cause and effect by which we can make valid predictions. Trivial existence at the whims of capricious and unknown forces is rejected, and we hope to find a stability that enables us to chart sensible courses into an ordered future extending from an equally ordered past.

Our view of the universe is a complicated process of observation, interpretation, and neural processing of signals from the senses. On the basis of these signals we individually construct an inner reality within the human brain that is unfortunately based on illusions generated by the nervous system on the basis of our individual experience and understanding. The picture of the universe that we see is not a single creation stored somewhere in each brain. Instead it is an evolving dynamic process of integrating information spread over many areas of the cortex. The result is an awesome range of human-generated universes each depending on the experience of each individual's exploration through his or her senses of the material world outside the human brain. Without exploration the human view of reality is extremely restricted, the individual personal universe is a microcosm instead of a macrocosm. Besides, without active physical exploration to stimulate the input to the senses while relying instead on passive inputs solely from packaged information supplied through television and books, cortical dynamic integration is distorted. Without giving our children an opportunity to explore, we shortchange them, and the human desire for exploration that starts in the cradle is stifled. The result is a falsely stimulated craving for a widened experience that is never satisfied by alcohol, drugs, and sexual excesses or other insatiable appetites so prevalent today on a global scale.

There were many exciting, horizon broadening discoveries from piloted lunar exploration, but this is common in most explorations. Terrestrial geologists can explore an area for years and still know that every day they are out in the field something new and unexpected may be discovered. Discovery is the very essence of exploration in all fields of human endeavor.

Many surprises about the Moon were regarding human misconceptions and armchair speculations that existed before the actual doing and observing. Recapturing the excitement of the early explorers of the American continent we find today equally as many unknowns and more unexplored fields than ever before in human history. There are challenging frontiers in medicine, biology, astronomy, nuclear physics, sociology, and economics. In almost every human endeavor we are confronted with unknowns and challenges. Far from closing the book about the Moon and stating that we now know all we need to know about that world, the Apollo science program proved again how little we really do know of reality and how much remains to be discovered. Lunar science forms a model on which we can thrust our curious minds at the other challenges, the other frontiers of humanity, as we direct the capabilities of reasoned thought to solve the many serious human problems confronting the people of the United States and of the world generally.

Although lunar science has made great strides from the speculations before the Apollo program, although we have now determined the ages of the most prominent lunar features and found incredibly old rocks dating back beyond those of

Earth, although we now have a much clearer view of how a planetary body such as the Moon or Earth evolved, we still are not clear about how and where the Moon originated. We still cannot say without reservation that it came from the Earth, or even from the same zone of the Solar System as the Earth, or if it originated elsewhere some 4.6 billion years ago in the extensive nebula from which the Sun and the planets formed.

But lunar science has allowed scientists to place boundaries on many theories as the result of accurate measurements made on lunar materials and in the environment of the Moon; measurements that provide precise information on ages of lunar rocks, and their chemistry, petrology, magnetic properties, information on the geology and geochemistry of the Moon's surface, on the seismology of its interior, and on the way the planet as a whole interacts with the medium of interplanetary space and the solar wind.

It is nearly impossible to define a planet from observations on a single continent, or define worlds from observation of a single world, or planetary systems from observations of a single system. Many data points are needed from various objects if one is to arrive at universally valid generalities. Additionally, as already mentioned, there are significant problems of defining reality from observations of matter alone; we need to investigate fields as well as particles over a wide range of energies, and radiation over an enormous spectrum.

We have to proceed step-by-step in the search to understand what is life, how did it originate, why is there consciousness, what is its purpose, how prevalent is life and consciousness in the material universe, and what is the role of consciousness? While exploration of the Moon is not likely to answer most of these questions it will provide a first stepping stone toward exploration of other parts of the Solar System in which answers, or partial answers, may be discovered when these parts can be explored through technology developed by a return to the Moon and the establishment of permanent outposts there.

Humans have always been bedeviled by major philosophical questions the answers to which have been sought by theologians delving into the mind of spirit and by explorers ranging beyond the limits of contemporary horizons: the great religious leaders, the major odysseys of the ancients documented in Greek literature, and the post-medieval navigators.

In the early 1970s I interviewed Nobel Laureate Hannes Alfven concerning humanity's space activities beyond Apollo. He had just returned from Stockholm where he had talked about the discoveries of the first and second decades of spaceflight and where we should go next with the newly developed capability to explore the physical world. "It is our moral and intellectual duty to describe the story [evolution of the universe and of the Solar System] in modern terms," he asserted.

He stated that trying to understand how Earth and the other planets formed is philosophically as important as trying to understand the structure of matter. Yet the latter task had absorbed most of the scientific research activity for the first two-thirds of the twentieth century. He expected a changed emphasis in physics with less concentration on the fundamental particles of matter and quantum mechanics and more research into how the Solar System and the planets formed from the original plasmas that pervaded the universe in the beginning. He was doubtful that exploration of the planets would provide all the answers because the surfaces of all the large bodies of the Solar System have been modified so much since their formation. Even the Moon's very ancient surface has been violently disturbed. But asteroids might provide important clues and they should be explored and studied, he concluded.

In this search for understanding, while the Moon may not give all the answers, outposts on its surface will provide new technology that will be essential to making exploratory missions to asteroids early in the twenty-first century. And the asteroids might provide information about the original plasma from which the Solar System formed.

Why not explore with machines alone? Why not send instrumented probes to find out the answers? As I wrote in my book *Assault on the Moon,* published by Hodder and Stoughton in 1966:

> Critics of the manned programme to reach the Moon have said that the whole thing could be done by machines. They are wrong. Machines are very good for relieving Man of physical and mental burdens but they are woefully inefficient as creative thinkers and questioning explorers.
>
> Man is much superior to remotely-controlled devices in lunar exploration. He is motivated while machines are not. A machine will fail when things go wrong. A man will attempt to put things right and carry on when it looks as though he should be giving up and failing like the machine. Man has better mobility and visual ability than any machine. He can manipulate complex tools and has very high reliability and versatility. His judgement is far beyond that of any machine, especially in complex unexpected situations.
>
> Even encumbered in a spacesuit a man can squeeze through crevices and holes and climb cliffs which would be impassable barriers for machines. This is because human limbs have three degrees of freedom, rapid data feedback, control couples with a power output that is infinitely variable, and are guided by sensitive multi-channel sensors. Man can also see stereoscopically, in colour, in any direction he chooses. He can focus his vision almost instantly from infinity to a few centimeters, and can adapt rapidly to light intensities ranging from full sunlight to faint starlight.

This generation has had the greatest human explorers of all time. We have leapt across nearly a quarter of a million miles of space to step on and unravel the secrets of another and quite separate and radically different world. We have developed the technology to expand into the Solar System, place humans on other more distant worlds and offer a challenge and an outlet for a generation of children of the space age. Today these children are frustrated internationally by the stifling environments of overly regulated and overly governed world societies of many political ideologies but with little if any long term leadership, and everywhere teetering on the brink of economic catastrophes.

A bold international leadership is needed to assist humankind into an expansion into the Solar System and to bring about a revitalization of the human spirit that can occur only as horizons of physical and mental awareness are expanded, the true nature of humans as inveterate explorers is recognized, and development of this nature is encouraged.

In 1989 President Bush asked the National Space Council to determine what was needed to chart a course into space, first with an outpost on the Moon and then on to a permanent human presence on Mars. This followed the lead of the recommendations of the 1986 National Committee on Space: ". . . to lead the exploration and development of the space frontier, advancing science, technology, and enterprise, and building institutions and systems that make accessible vast new resources and support human settlements beyond Earth orbit, from the highlands of the Moon to the plains of Mars."

At the July 20, 1989, anniversary of the first landing on the Moon, President Bush said:

> In 1961, it took a crisis—the space race—to speed things up. Today we do not have a crisis. We have an opportunity. To seize this opportunity, I am not proposing a 10–year plan like Apollo. I am proposing a long-range, continuing commitment.
>
> First, for the coming decade—for the 1990s—Space Station Freedom—our critical next step in all our space endeavors.
>
> And next—for the new century—back to the Moon. Back to the future. And this time, back to stay.
>
> And then—a journey into tomorrow—a journey to another planet—a manned mission to Mars.

On November 2, 1989, the President approved a national space policy that updates and reaffirms U.S. goals and activities in space:

> strengthen the security of the United States; obtain scientific, technological, and economic benefits; encourage private sector investment; promote international cooperative activities; maintain the freedom of space for all activities; and expand human presence and activity beyond Earth orbit into the Solar System.

In support of the initial directive, a NASA task force had conducted a 90–day study of a Human Exploration Initiative encompassing both robotic and human missions to achieve the objectives. This report was published in November 1989 also. Five reference approaches were presented to reflect the President's aims through an initial operation of space station Freedom, followed by a lunar outpost and then a manned landing on Mars and the ultimate establishment of a permanent human presence on the Moon and on Mars. The reference approaches were chosen as models to evaluate the effects of establishing various schedules and costs and other constraints or priorities. All concluded that a lunar base can be operational between the years 2008 and 2022, and a Mars base between 2018 and 2027, depending on the priorities and constraints chosen. The first humans to go back to the Moon would land between 1999 and 2007, and on Mars between 2011 and 2016. The study examined technical scenarios, science opportunities, required technologies, international considerations, institutional strengths and needs, and estimates of resources.

In mid-1991 a presidential panel confirmed these goals: return to the Moon by 2005, establish a permanent human base there as a precursor to landing Americans on Mars by 2014, and start utilizing the vast economic commons of the Solar System. Additionally, the panel recommended reviving the nuclear rocket program, especially for journeys to Mars. The report was primarily directed toward

EXPLORERS

19

technology feasibilities rather than cost projections. But even the most conservative cost estimates of the whole human exploration program extending over a quarter century is less than two years of the current defense budget to protect the nation from enemies who seem to have vanished almost overnight. At present the U.S. has fewer threats from overseas than at any time in its history and can easily redirect its efforts from major military expenditures to advanced technology investments that will have enormous economic multipliers and general benefits to society, as was experienced in the early days of the space program.

From the beginning of the U.S. involvement in space, and often with strong opposition by some scientists who feared funds would be diverted from their research, the space program has expanded human activities in the new environment beyond Earth's atmosphere. The national capability moved through Mercury and Gemini and on to Apollo and a landing of astronauts on the Moon. With Skylab it tested operations in low Earth orbit (LEO), an orbit just beyond the appreciable effects of the Earth's atmosphere, and followed with the space transportation system of the Shuttle. A space station, which will be vital to any major human expansion beyond Earth, is the next logical step. The Soviets, who generally were in the lead over the U.S. in space, had long recognized this and had people operating almost permanently in a space station environment for a number of years, while the U.S. fumbled along with its space transportation system and abandoned other ways of lofting humans into LEO and returning them to Earth in capsules. During 1992 the Russians offered the use of their MIR space stations to other nations, including the U.S.

The time has now arrived for action. Unless the U.S. as a nation clearly endorses the new space objectives and goes ahead with space station Freedom despite its undoubted limitations, the global space activities will move elsewhere from North America. To several politicians it is easy to attack the investment needed to develop the space station, key to the nation's future, and state that the money could better be spent to educate children or on any of the legion of social problems besetting the nation. However, the amount of national resources to be spent on Freedom is minuscule compared with such glaring wastes as the savings and loan debacle, the reckless abandonment of foreign loans, the continued testing of nuclear weapons, the maintenance of expensive military bases overseas, tobacco and other subsidies, and a seemingly endless list of federal and state government inefficiencies.

I talked with renowned anthropologist Margaret Mead in 1972 in connection with a lecture she was giving in San Francisco on our open-ended future. We must be wary of a tight either/or philosophy, she said. There is not the slightest proof that if we had not been working on space we would have done anything of greater human value. Mankind's future is in our own hands, in how we apply our own humanity. There is a broadening of our consciousness that started about the time the first Sputnik (Soviet artificial satellite) was launched. People are expanding their awareness of space and time in unprecedented ways. If the Apollo project had done nothing other than provide pictures of our planet as seen from the Moon, it would have been worthwhile because it made humanity see Earth for what it is; the fragile home of human beings. She stressed that one of the important questions awaiting answers was how to organize humankind to safeguard its future.

Human expansion beyond Earth is extremely important to the future development of our species. Freedom will allow humans to start building in space (figure

FIGURE 1.9 The first stage of the space age has been successfully completed with unmanned exploration of nearly all the planets of the Solar System and many of their satellites, landing of humans on another world, and long-term operations, by the Russians, in orbital space stations. The next stage of the human expansion is building in space and on other worlds as we have in the past built on Earth. In this illustration an astronaut is shown working in space out of the Space Shuttle cargo bay. (*NASA*)

1.9) as in the past they have built on Earth. Building in space and on other worlds will require that we develop entirely new technologies—those we have never used, or have needed to use, on Earth. The writer remembers how he received several relevant comments when some two decades ago he was interviewing engineers in a developing country about how U.S. know-how could be applied to help modernize the infrastructure of developing nations on the Pacific Rim and improve their economies. He was told that too often westerners tried to transfer technical knowledge based on textbooks generated for the European or North American environment. Dam building in tropical climates, for example, he was told, is entirely different from that in temperate climates; soils are different and the availability of material resources are different.

Yet all the textbooks were written for European conditions. We have to accept that the same will apply to operations in space and on other worlds. Civil engineering will have to develop entirely new technical disciplines for building in space and building on other worlds. Operations on the Moon and on Mars and future missions to mine asteroids will demand that large interplanetary transfer spacecraft be assembled, maintained, and refurbished in LEO, requiring innovative technology.

Freedom will explore this technology. It will also benefit from technical contributions by other nations, from modules to be designed and supplied by the Europeans and the Japanese. Both space groups are anxious to operate their modules in LEO and gain technological experience in the space environment. They will without doubt seek other ways to do so should the U.S. further drag its national feet and abandon or delay the space station program. Space station Freedom is needed to develop the specialized advanced technologies and logistics and control techniques for many rewarding space operations that are so essential for a united global humanity of the coming century.

2 THE MOON AS A WORLD

If you look at a Full Moon high in the midnight sky, you will—even if you are the most casual of observers—see obvious differences in brightness among its surface areas. The face of the Moon has two major landforms; smooth dark plains called maria, and lighter colored highlands called terrae. When viewed through even a small aperture telescope when the Moon is not at its full phase, the highlands are seen to be covered with overlapping craters of many sizes, while the dark plains are relatively free of these shoulder-to-shoulder craters but are textured with many small craters, a few recent impacts, and low ridges, and are generally surrounded by rings of mountainous terrain.

The Moon has always been important to humans. Before the invention of artificial lighting, Gibbous and Full Moons allowed night time activities that were otherwise difficult or impossible. The regular phases of the Moon permitted measurement of time in longer units than the day, and the period from one New Moon to the next became the measurement of the month and a vital element for the calendars of several major religions.

When people realized that the Moon is a world and not just a light in the sky, voyages to the Moon became popular in romantic fiction. Some ancient writings contain supernatural accounts of the Moon's hypothetical inhabitants, such as lost souls and demons. Others took a more realistic viewpoint. Democritus (450 B.C.), for example, said the Moon was a world similar to Earth with mountains and valleys. By 300 B.C. Aristarchus had calculated that the Moon was 224,000 miles (360,500 km) from Earth; amazingly close to its actual shortest distance, at perigee, of 225,732 miles (363,263 km). However, other writings repeat legends and fictional accounts in which the Moon is stated as being the abode of god-like beings. The Babylonian *Enuma Elish* mentions a Moon god, and ancient Egyptians recognized Khons and Thoth as male Moon-gods. To the Zidonians the Moon was the goddess Ashtoreth, the evil of whose worship is mentioned in the Old Testament.

In fiction as far back as ancient Greece there are accounts of contact with inhabitants of the Moon. Plutarch, writing early in the first century A.D. about the physical facts concerning the Moon in his *Faces in Orbe Lunare,* also mentioned his sorrow that inhabitants of the Moon would have unbearably longer days and nights, while souls in purgatory would ultimately be purified on the Moon.

About 100 years after Plutarch, Lucien wrote *Vera Historia,* in which a whirlwind carried a ship to the Moon. The narrative continued with a satire on the Greek wars in which the inhabitants of the Moon battled those of the Sun.

There are no records of writings about the Moon during Europe's Dark Ages, but in the seventeenth century Kepler wrote his novel *Somnium* in which the characters reached the Moon by demon power and there suffered various encounters with the inhabitants of the Moon. Many other fictional accounts followed. Francis Godwin used large birds to carry his hero to the Moon in his book *The Man in the Moon,* published posthumously in 1638. John Wilkins suggested use of wings, large birds, or some then uninvented engine to make a British national effort to colonize the Moon. His proposals, outlined in the book *Discovery of a New World* (1638), were supported by several members of the then newly formed Royal Society.

Voyage de Mond de Descartes (1691) by the French writer Gabriel Daniel employed a mystical method to get the hero to the Moon, similar to that which Edgar Rice Burroughs would use much later to transport his hero to Mars. The theme returned to the idea of souls on the Moon. Other writers used the Moon to satirize the chaotic nature of politics. Murdoch McDermott predated Verne by 150 years with his *Trip to the Moon* (Dublin 1728) in which a gunpowder-fired projectile carried the travelers there. Cyrano de Bergerac, in his *Histoire Comique des Etats et Empires de la Lune* (1656), used bottles of rising dew to propel his traveler to the Moon and then used an account of the lunar environment to satirize the book of Genesis.

The Consolidator by Daniel Defoe used a Rube Goldberg type of flying machine, described in confusing jargon, to carry his explorers to the Moon, and Edgar Allen Poe wrote a short story about the Moon, *Unparalleled Adventures of One Hans Pfall* (1835). Into the nineteenth and twentieth centuries novelists used mechanical methods to carry their heroes to the Moon. Jules Verne's *Around the Moon* (1867) reflected the age of great archeological discoveries by having his travelers catch intriguing glimpses of the ruins of ancient lunar cities as their spaceship circumnavigated the Moon before returning to Earth. H. G. Wells in his novel *First Men on the Moon* (1901) had his hero, Cavor, encounter an advanced race of selenites living deep beneath the lunar surface.

Only a few years later, in 1908, the Russian Konstantin E. Tsiolkovsky published one of the first serious papers on how rocket propulsion might be used to overcome Earth's gravity and travel into space. By 1920 the American engineer Robert H. Goddard published his short paper *A Method of Reaching Extreme Altitudes,* in which he showed how a multistep rocket could achieve sufficient final velocity to send a small charge of flash power to explode on the lunar surface and be visible from Earth. But Goddard backed off with the statement that this feat would be of little scientific importance though of undoubted general interest.

It remained for the German Hermann Oberth, in his book *Die Rakete zu den Planetenraumen* (1923) and a subsequent very much enlarged edition *Wege zur Raumschiffahrt* (1929), to establish in detail the technical requirements for a voyage to the Moon. He calculated that the size of a moonship would be equal to that of a small skyscraper (about the size of Apollo), which the savants of the 1920s regarded as being completely impractical. He also first suggested the use of atmospheric braking ellipses to slow the moonship on its return to Earth, now very much in vogue as new technology. Many serious books and articles followed Oberth's work and several in-depth studies were published, notably the report of the British Interplanetary Society's research committee in 1939. This committee was led by Ralph A. Smith and included Harry E. Ross, J. H. Edwards, A. Janser, M. K. Hanson, and Arthur C. Clarke (figures 1.1 and 2.1).

The study to establish the requirements needed to carry a crew of three men to the Moon and return them safely to Earth was documented in a series of articles in the society's Journal just before World War II. A design published in January 1939

FIGURE 2.1 A historic meeting in 1937 at Chingford, London, when Robert C. Truax, holding the small rocket motor, visited the British Interplanetary Society. Far left is J. H. Edwards, next to him Eric Burgess, then Harry Turner, also of the Manchester affiliates, then Ralph A. Smith and Maurice. K. Hanson, and finally Arthur C. Clarke. Edwards, Smith, Hanson, and Clarke were members of the committee that studied the cellular rocket as a means to carry three men to the Moon. (*Burgess archives*)

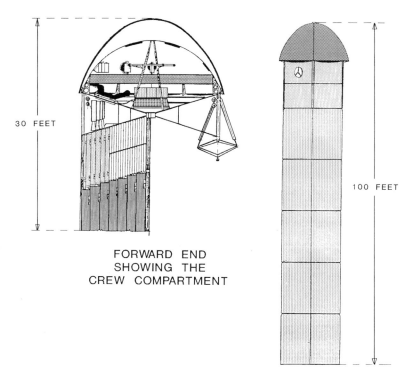

30 FEET

100 FEET

FORWARD END
SHOWING THE
CREW COMPARTMENT

FIGURE 2.2 One of the first in-depth studies of the requirements for a three-man voyage to the Moon was published by the British Interplanetary Society just before World War II. This illustration shows the crew compartment and the general form of the proposed moonship.

was a radical departure from all previous suggestions. It assumed a cellular construction of solid-propellant rockets (figure 2.2) in which spent cases would be jettisoned to obtain a very effective multistep design. The initial concept was for a launching from a site close to the equator to take advantage of the Earth's spin, and a direct journey to fly around the Moon and land back on Earth using parachutes as was, indeed, the design of the first Apollo mission. A liquid propellant lunar mission plan developed by the Manchester group of British interplanetary enthusiasts about the same period used a rotor blade Earth landing system. Later studies by Smith and his team developed suggested engineering approaches to land on the Moon using vehicles somewhat similar to the Lunar Excursion Module of Apollo. In 1952 in his book *Rocket Propulsion* (London: Chapman and Hall), the writer suggested use of assembly of a mission to the Moon in low Earth orbit (LEO). It was not used for Apollo because the technology of operating in orbit was still not developed, but this method is now in vogue for the next stage of manned exploration of the Moon using a transportation node in LEO, as is described in a subsequent chapter.

The space age that began with the 1957 orbiting of the Russian Sputnik I brought the realization that the Earth cannot be fully understood without reference to the other planetary bodies of the Solar System. Each planetary body differs in significant ways from the others, although all may have had a common origin. Big questions were how the planets formed and what caused them to evolve so differently to reach their present states. The more distant bodies from the Sun are gas and ice worlds; outer planets are gaseous giants with possibly small rocky cores, and some of their icy satellites rival in sizes those of the inner planets. Comparative planetology applies observations of the different planets and satellites to try to find common circumstances. Scientists thought that the Moon, for example, could show evidence of the primitive composition of the Solar System if the Moon's surface had not been changed radically since its formation. Mars might supply information about how life evolves from prebiotic molecules if its climate were more Earth-like soon after the planet's formation. Venus might give information about how a runaway greenhouse effect could destroy a planet's oceans and raise its surface to a temperature at which life could no longer exist.

An intriguing question before spacecraft reached the Moon was whether it was completely dead or still had seismic activity with tectonic movements and vulcanism. Some astronomers observing the Moon from Earth claimed they had seen evidence of emissions of gas and bright glows on the Moon's surface that might be caused by current volcanic activity. They had also identified many fracture patterns indicating that the crust had been molded by internal forces as well as by impacts.

The diameter of the Moon is 2159 miles (3474 km). As viewed from the Earth's surface, the Moon has a mean angular diameter of 31 minutes 7 seconds, which is about the same as the much more distant Sun. Since the Moon orbits the Earth in an ellipse, its distance from Earth varies from 226,000 miles (364,000 km) to 252,000 miles (405,547 km) so that its apparent diameter varies from 29 minutes 20 seconds to a maximum of 33 minutes 36 seconds of arc.

At an average orbital speed of 0.63 miles per second (1.0 kps) the Moon takes 27 days 7 hours and 43.192 minutes (27.321661 days) to make one complete revolution around the Earth relative to the stars. This is the sidereal month. Because the Earth and Moon move together around the Sun, the period from one phase of the Moon to the same phase; e.g., Full Moon to Full Moon, or New Moon to New Moon, is about two days longer than the sidereal period. This phase-to-phase period is the synodic month, also called a lunation. It has a mean period of 29 days, 12 hours, 44.05 minutes (29.530588 days).

As the Moon moves around the Earth over its lunation its sunlit hemisphere is viewed at different angles from the Earth resulting in the phases (figure 2.3). After the New Moon, which astronomically is when the Moon is between Sun and Earth, the phase is a waxing (increasing) thin crescent. The first appearance of the thin crescent is recognized as the first day of the month in some religions. The phase continues to increase to a half Moon (First Quarter) and then a waxing gibbous Moon leads to the Full Moon when the Moon is on the opposite side of the Earth to the Sun. After the full phase, the waning (reducing) gibbous Moon leads to the Last Quarter half Moon followed by a waning crescent and back to the New Moon.

It is sometimes amazing how artists forget about these principles of the Moon's motions and phases. I remember how not so long ago I received an urgent request from India from a friend working on the motion picture *Passage to India*. We had earlier worked together at Pinewood Studios, London, on the James Bond movie, *Moonraker*. Apparently the director, David Lean, needed a good photograph of the Moon for inclusion in one scene. I was able to put my friend in touch with Lick Observatory for the Full Moon photograph the director required. Imagine my

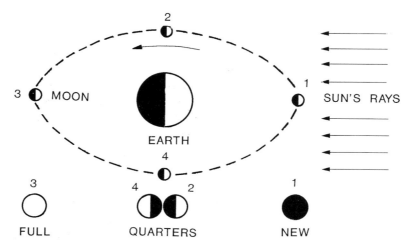

FIGURE 2.3 The Moon presents phases to observers on Earth as the relative positions of Moon, Sun, and Earth change during each lunation as shown in this diagram.

chagrin when I viewed the complete motion picture and saw the camera reveal a Full Moon over the stark cliffs of India. Everything looked perfect except the action was taking place close to midday with a Full Moon riding high in the sky! A Full Moon is always opposite to the Sun and must be below the horizon at midday. Also many artists use a crescent Moon to indicate it is late at night, whereas crescent moons are visible only just after sunset or just before dawn with the horns of the crescent pointing away from where the Sun is below the horizon. This seems also to be a problem with fiction writers, who have Full Moons rising at midnight and have sunrises and sunsets at the wrong times of the year for the times of day they are depicting.

The actual length of a lunation varies from the mean value of 29.530588 days because both the lunar and terrestrial orbits are elliptical. For example, if the Moon comes between Earth and Sun (New Moon) close to when the Moon is nearest to Earth (perigee) the lunar month is slightly shorter than average. If the conjunction occurs close to lunar apogee (farthest from Earth) the lunar month is longer than average. As for the Earth's elliptical orbit, lunar months are slightly longer when Earth is near perihelion (closest to the Sun), and shorter when it is near aphelion (most distant from the Sun). At present, the longest months occur in January but will gradually move into February during the next 200 to 300 years. The shortest months occur six months from the longest months.

F. Richard Stephenson and Liu Baolin have reported in *The Observatory* (February 1991) that based on computations covering New and Full Moons over a period of 5000 years, the lunar month has varied from a minimum of 29.2679 days (in 302 B.C.) to a maximum of 29.8376 days (in 400 B.C.). Now the maximum and minimum periods are 29.83 and 29.30 days respectively, and both are tending toward the mean value.

The result is that religious calendars that rely on lunations and use alternate 29–day and 30–day months to average out the 29.5306–day cycle can periodically be more than one day in error. The Chinese calendar, which uses other sequences of

29-day and 30-day months, closely approximates the lunar motions and keeps the error to within one day.

When Galileo applied the newly invented telescope to the Moon early in the seventeenth century, he recounted how he saw craters and mountains. His first drawing carries a date of 1610. It was made at the time of First Quarter and shows the Lunar Alps and Lunar Apennine mountain ranges and several large craters one of which was probably Clavius (see figure 2.4). This is the first known drawing of details on the Moon.

About 40 years afterward, a 12-inch (30-cm) map of the Moon was drawn by Hevelius who, like Galileo, thought that the dark areas were oceans. He named them maria and surrounded them with seas, bays, and lakes, which he also named, although not with the names used today. He identified several bright ray craters. The present names of the maria first appeared on a chart produced by Riccioli in 1651. His names for the bright highland areas did not, however, survive, although his names of prominent craters did; Ptolemaeus, Copernicus, Theophilus, Grimaldi, and Tycho, for example.

In the following 300 years many selenographers produced maps of the Moon in increasing detail and with excellent accuracy despite the small apertures of the telescopes they had to use. The major advances came in the eighteenth century when Johann Hieronymus Schroeter built an observatory at Lilienthal. He used a 19-inch (48-cm) aperture Schrader telescope and another telescope built by Herschel to map the lunar surface. Although he never produced a map of the entire surface visible from Earth he made many detailed drawings (figure 2.5) and published a monumental work, *Selenotopographische Fragmente,* in 1791. He believed that intelligent beings inhabited the Moon and that it was an active world with volcanoes and an atmosphere.

Unfortunately most of his work and original observing records were lost when Napoleon Bonaparte's French invaders set fire to his two observatories and completely destroyed them.

Another expert observer was born in Berlin in 1794. Johann Heinrich von Madler teamed with a wealthy banker Wilhelm Beer and in 1824 started a geometrical

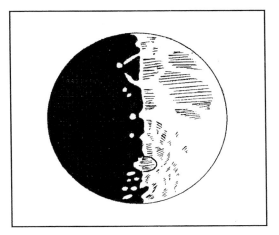

FIGURE 2.4 One of the earliest known drawings of the surface of the Moon was a sketch of the First Quarter made by Galileo in 1610, showing craters, mountains, and maria.

FIGURE 2.5 Schroeter's book, *Selenotopographische Fragmente,* published in 1791, contained plates such as this which shows part of Mare Imbrium. (*Manchester Central Library*)

survey of points on the lunar surface. With a telescope of only 3.75–inches (9.5–cm) aperture they catalogued the positions of 919 points on the surface of the Moon. They measured the heights of more than 1000 lunar mountains from the lengths of their shadows, ascertaining that some peaks were more than 23,000 feet (7000 m) high. They measured the heights of crater walls over crater floors, discovering, for example, that Tycho's walls rise nearly 19,000 feet (5800 m) above the crater's floor.

Beer and Madler published a chart of the Moon that remained the standard for half a century. Also they published in 1837 a book, *Der Mond,* which contrary to the opinions of their predecessors regarded the Moon as a dead world.

The next lunar chart of significance was the 40–year effort of Johann Friedrich Julius Schmidt during which he measured the positions of 4000 objects and the heights of lunar mountains and depths of craters. He produced a chart in 25 sections at a scale of 6 feet (2 m) for the lunar diameter and showing 30,000 objects on the visible surface. The chart itself took six years to prepare for publication in 1879.

Schmidt believed he had observed changes on the surface of the Moon during many years of observations. He claimed that one crater, Linne, had disappeared. He raised the question of whether or not the Moon was a completely dead world as had been suggested by Beer and Madler.

During the early years of the twentieth century even more detailed charts of the Moon were produced. Walter Goodacre published a lunar map of 72–inches (183–

cm) diameter in 1910. H. P. Wilkins produced a 60–inch (150–cm) diameter map in 1924 and a 300–inch (760–cm) diameter map in the late 1940s.

However, the accent changed to use of photographs to record details of the lunar surface more accurately than drawings could be. As early as 1862, Warren de la Rue received the coveted Gold Medal of the Royal Astronomical Society, London, for his lunar photographs. In France Maurice Loewy and his assistant Pierre Puiseux began photographing small sections of the Moon at the Paris Observatory. Their work culminated in the *Paris Lunar Atlas,* published in 1897. The Harvard astronomer William Henry Pickering obtained 80 excellent plates from his work in Jamaica. These images were included in his book published in 1903.

In the twentieth century the outstanding work for many years was a series of plates taken with the 100–inch (2.5–m) telescope at Mt. Wilson Observatory, California, followed later by excellent images obtained at Lick and Pic du Midi observatories.

A quantum jump in mapping the Moon came, however, just after mid-century when it became apparent that spaceflight was inevitable and that people would one day soon visit the Moon. The U.S. Air Force started a program to produce a large-scale, accurately oriented map of the Earth-facing hemisphere of the Moon. In 1960 the *U.S. Air Force Lunar Atlas* was published—a comprehensive selection of the finest lunar photographs available at that time. The 280 images in the atlas originated from Mt. Wilson, Lick, Pic du Midi, MacDonald, and Yerkes observatories. Each of 44 fields into which the Earth-facing hemisphere was divided was shown by at least four photographs taken under different angles of illumination. This provided enhanced details of topography in those images where the sunlight cast long shadows, and albedo details in images where the sun shone directly down on the lunar surface. In 1961 a supplement appeared with a selenographic grid plotted at intervals of 0.01 of the lunar mean radius.

Another supplement, entitled *The Rectified Lunar Atlas,* published in 1965, contained photographs that had been rectified by projecting them onto a white hemisphere and rephotographing the projected image with the camera oriented vertically over the point of interest. This was done to remove the foreshortening of objects close to the visible limb of the Moon. Thus craters that appeared elliptical close to the limb regions were rendered in their true shape on the rectified pictures. All the earlier lunar maps had shown the Moon as a globe, the way that astronomers were used to seeing it through a telescope. A new set of U.S. Air Force charts were projected similar to the maps of Earth's regions we are familiar with seeing. Each map was a rectangular chart with contour lines and graduated shading to show features (figure 2.6a). The names of mountains, craters, plains, and valleys were given. The whole of the visible face of the Moon was covered by 144 separate charts. More detailed charts covered the equatorial regions where the Apollo astronauts were expected to land. These charts were on a scale of eight miles to the inch. Later, even more detailed charts of expected landing sites were made on a scale of a half mile to the inch.

The major charting problem for Apollo was the conversion of Earth-observed control points of relative positions of landmarks, such as craters, into lunar-centered latitude and longitude that could be used for navigation to a successful landing. Data from the Lunar Orbiter photographs were used to do this by relating time to

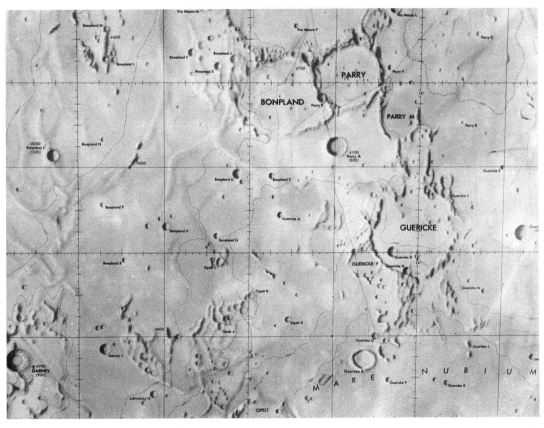

FIGURE 2.6 (a) Part of the U.S. Air Force chart of the Moon showing the region west of Guericke where Ranger VII photographed the surface as it plunged down to a hard landing. The area was named Mare Cognitum after the Ranger mission, meaning the "Sea that is now known." (*U.S. Air Force Chart and Information Center*)

orbital position. The result was a set of descent monitoring charts to guide the Apollo astronauts to a point at which radar altimeter guidance could be used for the final touchdown. These charts were first checked out on the Apollo 10 mission and used for the landing of Apollo 11.

Meanwhile, the U.S. Geological Survey, in a joint program with the Army Map Service, produced a general photogeologic map of the Moon and then a series of geologic maps to a scale of 1 in 1,000,000 (figure 2.6b). The geology of much of the Earth-facing side was derived from interpretation of telescopic photographs and visual observations at various telescopes. Geological units were identified on the basis of relative age by application of the principles of stratigraphy; i.e., older units

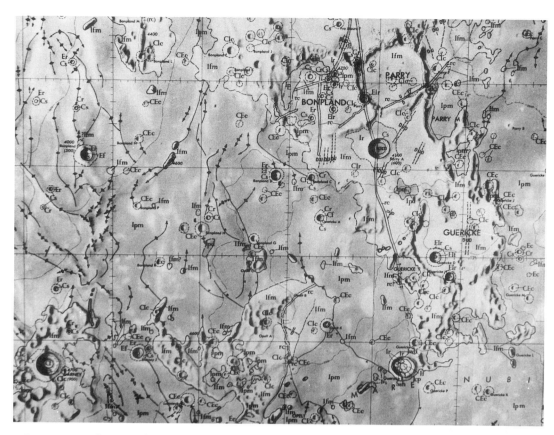

FIGURE 2.6   (b) A geological map of the same area. (*U.S. Geological Survey*)

being overlain and modified by younger units. Thus geologists established the relative ages of the various features on the Earth–facing hemisphere (figure 2.7).

This process of relative age-dating placed the oldest geological units as the heavily cratered highlands that existed before the impacts that created the mare basins. The next period was the formation of the Imbrium unit when Mare Imbrium and the other large mare basins formed, followed by Eratosthenian units and Copernican units. These most recent geological units were named after the major craters associated with prominent versions of the units. All the pre-Imbrium, ancient areas of the lunar surface are subdued in their topography and heavily modified by overlapping cratering. During the Imbrium period several major circular basins, each hundreds of miles across, were formed by the impact of large bodies coming from space. The basins were subsequently flooded by lava flows.

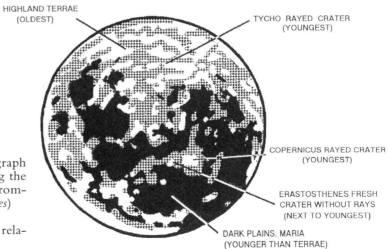

HIGHLAND TERRAE
(OLDEST)

TYCHO RAYED CRATER
(YOUNGEST)

COPERNICUS RAYED CRATER
(YOUNGEST)

ERASTOSTHENES FRESH
CRATER WITHOUT RAYS
(NEXT TO YOUNGEST)

DARK PLAINS, MARIA
(YOUNGER THAN TERRAE)

FIGURE 2.7   (a) Photograph of the Full Moon showing the light and dark areas and prominent craters. (*NASA-Ames*)

(b) Drawing to illustrate relative ages of lunar features.

The Earth-facing hemisphere was changed considerably by the formation of the Imbrium basin. The impact showered the surface of the Moon with projectiles that in turn gouged out other craters. Waves of pulverized material sloshed like liquid across the lunar surface filling many valleys and ancient craters. A vast circle of mountains was raised by the impact and formed the boundary of the basin.

Included in the Imbrium age are extensive areas of basaltic lava that flooded the basins after their original formation by impact. This flooding appears to have stretched over a considerable time; it was most probably episodic rather than continuous.

Craters on the Imbrium material that do not have bright ray systems form the Eratosthenian unit. Later impacts produced sharp craters surrounded by bright rays of ejecta. These are the features of the Copernican period that extends to the present.

Without landing on the Moon and sampling the lunar materials it was not possible to establish absolute ages for the different units. Since the Earth's surface has changed almost entirely during the last few hundreds of millions of years, no clues could be derived from the terrestrial record. There was no way to decide when the large bodies impacted on the Moon; whether millions or billions of years ago.

When, how, and where the Moon originated was the subject of many speculative theories. From examining records of ancient eclipses it was found that the angular momentum between Earth and Moon is being redistributed between the two bodies. Tidal friction has caused the rotation of the Earth to slow and the Moon's distance from Earth to increase. The discovery of these effects led to speculations, based on an assumption that the effects had continued at a constant rate, that the Moon had been within about 62,000 miles (100,000 km) of Earth only 1.5 million years ago. Some astronomers speculated that the Moon formed as a chunk of Earth spun off from the part that is now the Pacific Ocean. Other speculations were that the Earth was spinning very rapidly early in its several billion year history and the outer layers spun off and led to the formation of the Moon. Some computer simulations suggested that the Moon was closest to the Earth several million years ago but before then it was further away. This resulted in speculations that the Moon had formed elsewhere in the Solar System and was captured by Earth. Another speculative theory was that the Moon and Earth formed together as a double planet as they accreted from condensing material of the primeval solar nebula. In the last decade a theory has gained popularity, according to which the Moon originated as a result of the cataclysmic collision of a body the size of Mars with the developing Earth.

Formation as a double planet presents problems. Earth and Moon have quite different densities, and they would be expected to be similar if the two bodies formed from the same group of planetesimals; also the orbit of the Moon would be expected to be closer to the plane of the ecliptic than it is.

As for the origin by spin off from Earth, this theory also suffers from the inclination of the Moon's orbit differing from the Earth's equatorial plane and the total angular momentum of the system being insufficient. The third speculative theory of the Moon's origin, capture by Earth, satisfies many dynamical considerations associated with the present Earth/Moon system. Moreover, there are other

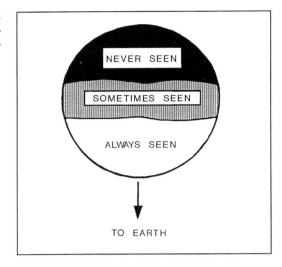

FIGURE 2.8 Principally because the Moon's orbit is not circular, but elliptical, lunar librations let us see from Earth more than 50 percent of the lunar surface.

instances of large satellites equal to or greater in size than the Moon. One supporting example is that of Triton whose highly inclined, retrograde orbit around Neptune suggests that it is a captured satellite. Large as well as small satellites might therefore be captured by planets. The capture theory also might account for the big mare basins, because capture of a big satellite would be expected to sweep up by collision many smaller satellites. Collisions with such small satellites might have been the mechanism for producing the big basins on the Moon, thus placing its capture as taking place about 4 billion years ago.

There are other more recent theories for the formation of the Moon. In one theory the Moon formed from a ring of debris around the Earth. This debris came from colliding planetesimals that approached close to Earth but did not impact on its surface. Instead they broke apart on their close approach, which allowed some to enter orbit. Another theory is that planetesimals in orbit around the Earth, which formed simultaneously with Earth, accreted to form the Moon in orbit around the Earth.

In whatever way the Moon formed or was captured the process must have had important and most probably very dramatic effects on the evolution of the early Earth. The question of the origin of the Moon might be answered only when we have fully explored it and compared its bulk composition with that of the terrestrial mantle. On the basis of the Apollo results these two compositions appear to be different, but more data are required to eliminate a number of uncertainties. Also, we need to obtain more detailed information on the other satellites and planets of the Solar System, especially the large Moon-sized satellites of the outer planets, to help solve the formation puzzle. Observations of the Moon's motions as seen from Earth, particularly the way in which it oscillates so as not to present always exactly the same face to Earth (figure 2.8), enabled mathematicians to determine that the Moon possesses unequal moments of inertia; each a measure of the resistance of the Moon to changes in spin rate about an axis. An explanation was that the Moon bulges toward the Earth, but whether this was an actual physical bulge or merely a difference in density of material within the Moon could not be determined from Earth.

A big question was how such a bulge might originate. Was it a frozen tidal bulge, induced in a plastic Moon by the Earth very early in lunar history? Was it a bulge produced by convection from a hot interior oriented toward the Earth? Was it an accidental distortion of the Moon as it cooled?

As is discussed in more detail in a later chapter, the space program provided some clues but not all the answers. Tracking data from Lunar Orbiter spacecraft provided the first new evidence about the complexities of the Moon's gravitational field. Despite limitations because the Lunar Orbiter spacecraft could not be tracked when behind the Moon, the experiments showed that gravitationally the Moon is not a simple homogeneous body.

The most accurate measurements of the shape of the Moon were obtained by radar and laser altimeters carried in the Apollo missions. These instruments measured the precise altitude of several Command/Service Modules as they orbited the Moon. The best fitting circle about the Moon has a radius of 1079.8 miles (1737.7 km). The dark maria surfaces are extremely level, sloping less than one tenth of a degree over hundreds of miles. Generally they are 2.5 miles (4 km) below the mean altitude of the highland areas. Instead of a physical bulge toward Earth, the Moon's Earth-facing hemisphere is flattened, while the far side is high. However, the center of mass of the Moon is displaced toward the Earth side of the Moon, along 24 degrees east longitude, about 1.5 miles (2.4 km) from the center of the geometrical figure.

Almost all lunar craters are now recognized as impact craters and most were formed very early in the history of the Moon. Large objects falling from space made craters up to 600 miles (1000 km) across. The smallest objects, mere specks of dust traveling at high velocity, made microscopic craters less than one millionth of a meter across, continually eroding the surface of rocks. The impacts that produced major craters not only spread lunar material over vast areas of surface to form layers of ejecta, but also showered debris in pieces large enough to form many secondary craters.

The number of craters of each size suggests there might have been two distinct periods of cratering; the tail end of accretion if the Moon formed from bodies condensed from the solar nebula, and a later period of cataclysmic bombardment by another group of bodies. During this second bombardment period the major basins of the maria were excavated. Since that period, cratering has continued but at a much reduced rate.

Collisions have been an important factor in the evolution of all bodies of the Solar System. All those with solid surfaces have been heavily cratered similar to the Moon. Where the surfaces have not subsequently been molded by tectonics or by weathering, the evidence of these early bombardments is preserved on the surfaces today, not only on the bodies of the inner Solar System but also on the satellites of the large outer planets. Images of the asteroid Gaspra, obtained in 1991 by the Galileo spacecraft on its circuitous voyage to Jupiter, show that it, too, is cratered much like the small satellites of Mars, which are believed to be captured asteroids.

The rings of the outer planets are probably also the debris from collisions, and the small satellites of the outer planets and of Mars and many asteroids might be the shattered remains of larger bodies that were fragmented by collisions.

There is a growing appreciation of what cratering can do to the surface of a planetary body, including Earth. It mixes the rocks, transports materials over vast distances, erodes earlier features, melts rocks, generates surges of material that behave almost like liquid and 'flow' across the surface filling valleys and smothering mountains. Evidence mounts that the early history of planetary bodies in the Solar System including the Earth was characterized by intense and devastating bombardments from interplanetary space.

The presence of cometary bodies and Earth-orbit-crossing asteroids could lead to a major impact on Earth at any time. Many close approaches have been observed. One small asteroid ricocheted off the top of Earth's atmosphere and there is evidence of many craters of various ages gouged out on the Earth's surface.

The impact of such bodies on Earth may have been responsible for major climatic changes and the extinction of living things. The ocean would retain the heat energy from impacts much longer than land surfaces. Impact in an ocean would have sent tremendous quantities of water vapor into the atmosphere. Impacts on land would fill the atmosphere with dust. Both would lead to major climate changes and cataclysmic effects on the biosphere. David Morrison has calculated that major impacts occur on Earth every 300,000 years on the average. A most recent impact is the one that formed the 0.75–miles (1.2–km) diameter, 600–feet (180–m) deep, Barringer meteor crater (figure 2.9) in the desert near Winslow, Arizona. This impact occurred about 50,000 years ago. Another important major impact is thought to have been where the Yucatan Peninsula is nowadays. This impact spread iridium and glasses for hundreds of miles around the impact site and is currently thought by some scientists to be the cataclysmic event that wiped out the dinosaurs and many other terrestrial life forms at that time. An incoming body of about one mile diameter would pack the explosive power of millions of nuclear bombs. Its crater would be at least 10 miles in diameter and its effects would be felt planetwide.

On the Moon, where the evidence of such impacts has been preserved over billions of years, the largest lunar craters are mountain-ringed mare basins hundreds

FIGURE 2.9 All planetary bodies suffer bombardment from space. The most recent event on Earth blasted this huge crater in the desert near Winslow Arizona about 50,000 years ago. The crater is 4,150 feet (1,265 m) in diameter. (*Photo by author*)

FIGURE 2.10   This unusual view of a Full Moon was obtained by Apollo 11 astronauts some 10,000 miles (16,000 km) from the Moon on their homeward journey. The large circular basin in the center is Mare Crisium. (*NASA*)

of miles across (figure 2.10). These lava filled basins are prominent on the Earth-facing hemisphere of the Moon. Several basins on the far side are not flooded with dark basaltic flows and are somewhat obscured by jumbled masses of overlapping craters. Recently a very large basin was discovered on the far side near to the south polar region. It appears on images of the Moon obtained by Galileo as the space-craft, after a gravity-assist from Venus, sped by the Earth/Moon system on its multiplanetary flyby journey to Jupiter. The Aitken Basin is more than twice the diameter of the prominent Crisium and Humorum basins on the Earth-facing hemisphere. It is ringed by concentric mountains and appears older and more degraded than the Orientale basin on the far side. It, too, is not filled with lava flows like the basins on the near side.

When an asteroid-sized body hit the Moon, rock was excavated from deep in the crust and ejected to produce a huge crater. Some excavated fragments fell back and dug secondary craters; others received so much energy that they escaped completely from the Moon. Some even hit the Earth. However, most of the debris that was excavated fell back to the Moon's surface near to the primary impact site and generated a cloud of material that spread across the Moon, a rolling mass of debris following an expanding curtain of projectiles gouging the surface at increasing distances from the main crater. Many secondary craters were buried by flowing debris; others were formed afterward from projectiles that had followed long circumlunar trajectories before returning to crash explosively into the surface.

The walls of the main crater collapsed inward. Its center rebounded and uplifted into a central area of mountains (figure 2.11). Rocks melted by the energy of impact flowed to form lava pools in the crater. Inward slippage of the crater walls produced concentric rings of mountains and high escarpments facing the basin. Pools of melted rock collected in valleys between mountains. Mare Orientale (figure 2.12) is an example of this type of young basin.

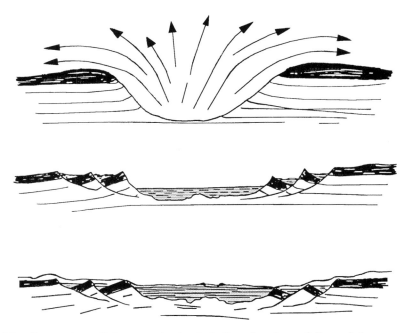

FIGURE 2.11   Formation of the mare basins is believed to have followed the sequence shown in this diagram. The major impact blasted a great hole, tilted rock strata, and spread debris over surrounding terrain for hundreds of miles. Rocks melted under the heat released by the energy of impact, walls slumped, and lava flowed into the basin, partially filling it. Much later, radioactive elements in the blanket of debris surrounding the partially filled basin released heat sufficient to melt rocks, giving rise to lava that flowed through cracks and fissures into the basin adding additional layers, and developing wrinkle ridges and collapsed lava tubes as it cooled. During the following quiescent period the relatively smooth lava flows were broken by a diminishing number of impacts and covered with a rubble of debris.

FIGURE 2.12   Mare Orientale at the center of this Galileo photograph of much of the lunar farside is believed to be the youngest impact basin. It was blasted out later in lunar history when sufficient radioactive heating could no longer be generated to flood the basin with lava. However, it appears to have formed before the flooding of the other big basins because none of its ejecta can be seen crossing lava flows on those basins. The wide dark expanse of Oceanus Procellarum is top right and Mare Humorum is the circular lava filled plain at lower right. At lower left is the vast Aitken unfilled basin first discovered on the Galileo pictures. It has twice the diameter of the Orientale Basin and is assumed to be older because of its degradation by impact cratering. (*NASA/JPL*)

FIGURE 2.13   The plains of lava flows were the initial landing sites for manned missions to the Moon. But even these plains are speckled with craters down to microscopic sizes. This view shows a typical basin surface with wrinkle ridges and various flow fronts. The prominent crater in the foreground is Bruce. It is about 3.7 miles (6 km) in diameter. (*NASA*)

Plains of basalt form the second major lunar landform. These fill the mare basins (figure 2.13). The basaltic layers on lunar maria are richer in iron oxide and deficient in alkalis compared with terrestrial basalts. They fall into four distinct groups: high titanium and low alkali and aluminum; high titanium and alkali and low aluminum; low titanium, alkali, and aluminum; and low titanium and alkali and high aluminum. Precious metals (siderophiles) are depleted in all the lunar basalts.

One of the unusual findings from the Apollo expeditions to the mare plains is that the concentrations of volatile and siderophile elements in the soil of the plains are higher by up to one hundred times those in the underlying mare basalts. While basalts might have been released from the highland rocks by the heat of impacts and then condensed in mare areas, the most likely source for the siderophile components is the meteorites themselves. Their relative abundance is similar to that in carbonaceous chondrites and different from that in lunar highland rocks sampled by Apollo.

Lunar mountains (figures 2.14 and 2.15) result from cratering and basin-forming processes; there are no lunar folded mountains like those on Earth. When a large body impacted the Moon the crust around the impact site tilted upward and formed an escarpment. Subsequent slippage of material toward the crater cavity produced concentric mountain rings.

Some smaller craters also have inner rings and central peaks. Central peaks are present in craters more frequently when the diameter exceeds 6 miles (10 km). The

FIGURE 2.14 Lunar mountains are not folded like terrestrial mountains but result from the basin-forming processes tilting strata. This image shows three major lunar ranges; the Lunar Apennine (top), Caucasus Mountains, and Lunar Alps. The large crater near the bottom of the photo with its floor deep in shadow is Plato. To its left is the Alpine Valley. The Hadley Rille area is in the upper left-hand corner. (*Jodrell Bank photo by author*)

FIGURE 2.15 The tilted strata of a lunar mountain are shown in this Apollo 15 photograph of Silver Spur in the Hadley Delta region at the Hadley-Apennine landing site. The upper strata are each 200 feet (60 m) deep. Below them are much thinner layers. This layering probably originated from successive flows of lava. (*NASA*)

central peaks probably formed by rebound after the impact. With increasing crater size there is often a ring of peaks between the center and the crater wall. Craters exceeding 90 miles (140 km) diameter often have multi-ring structures similar to the large basins.

The closest lunar feature to terrestrial mountains are low-lying hills, known as wrinkle ridges, on the mare surfaces. Since the ridges are concentric with major basins it seems that they are the result of lava flowing through circular ring-faults surrounding the original basins.

A relatively small aperture telescope reveals lunar rilles. Apollo 15 landed alongside one, the Hadley Rille in Palus Putredinis at the foot of the Lunar Apennine Mountains (figure 2.16). Rilles are of two types; fault troughs and sinuous channels. Some fault troughs are straight rilles that cut across other surface features and run for hundreds of miles. They are usually about 1.2 to 3 miles (2 to 5 km) wide. Others are curved. The troughs have flat floors and cliff-like walls. Usually they are located close to the edges of mare basins.

Sinuous channels look like erosional channels rather than faults (figure 2.17). At one time some astronomers suggested that these meandering rilles showed that water had once flowed on the surface of the Moon. However, as mentioned previously, samples of surface materials from the Moon show that the Moon never possessed significant amounts of water. Since the material that fills the maria is basaltic lava, all the sinuous rilles are most probably lava channels. The sinuous rille visited by Apollo 15 astronauts (figure 2.18) averages 0.6 miles (1 km) in width and

FIGURE 2.16 Hadley Rille is a prominent feature, even in a small telescope, in the Palus Putredinis area at the foot of the Apennine mountains. Note also the linear fault running diagonally across the area and the parallelism of the low range of hills to the fault. (*NASA*)

FIGURE 2.17 Sinuous channels are thought to be lava channels eroded by different episodes of volcanic eruptions of very fluid lavas (*NASA*).

FIGURE 2.18 The sinuous rille visited by Apollo 15 astronauts has the features of a lava flow: V-shaped and deepest at its widest parts. (*NASA*)

meanders from a cleft in the base of the Apennine Mountains. It is V-shaped and is deepest at its wider points, a feature of lava flow as opposed to channels formed by flows of water.

Before the landing of spacecraft and humans on the Moon there were many speculations concerning the surface and whether it could support a spacecraft or a human. Infrared and microwave observations from Earth indicated that fine dust covered the surface to a depth of several inches below which was a thick layer of gravel-like material extending to a depth of at least 60 feet (20 meters). Soviet radio work suggested that the surface was porous to a depth of several yards (meters). With radar pulses bounced off it the Moon behaved more like a polished sphere than it does in visible light. Indeed the reflected light from the Moon is peculiar as there is no appreciable darkening of the limb regions. This optical effect, first noticed by Galileo, results from the lunar surface's generally reflecting light back toward the source, i.e., back toward the Sun like a beaded screen used to project images of transparencies. Speculative theories attributed this back reflection to a delicate structure of vacuum sintered lunar dust or to a surface of multiple cracks and fissures. Another suggestion was that multifaceted particles like snowflakes covered the surface. Other speculations were that the fine dust would create pockets into which any spacecraft or person would be engulfed by the moondust.

What was discovered is that the surface is, indeed, porous, has a complicated fine structure, has many glass beads, and has shallow depressions down to microscopic sizes caused by the continued rain of bodies of a wide range of sizes. Moreover, the spacecraft determined that the surface is substantial enough to support large space-craft such as the Lunar Excursion Module of Apollo and for wheeled vehicles to travel safely and for humans to walk safely across it.

3 ASSAULT ON THE MOON

When the technology had been mastered for establishing artificial satellites in orbit around the Earth, the tantalizing proximity of the world of the Moon with its spectacular scenery offered a challenge that could not be ignored. Engineers and scientists had new tools of exploration forged from the era of development of long-range ballistic missiles based on the German technology that had produced the V-2 long-range ballistic rocket. In Soviet Russia and in the U.S., German engineers and scientists shared their expertise to help develop more powerful ballistic rockets, and when the military needs had been satisfied activity changed to the conversion of the big rockets into launch vehicles for spaceflight: a major challenge to humankind's spirit of adventure and exploration.

There were concurrent efforts with early satellites and lunar probes; and exploratory sorties reaching for the Moon with unmanned spacecraft: Pioneer, Lunik, Ranger, and Surveyor; and a certain amount of competition between sending men or machines into space, and who would be first to land on the Moon. Would it be the automated spacecraft being developed at the Jet Propulsion Laboratory, Pasadena, or the manned spacecraft being developed at the Manned Spacecraft Center, Houston, later named Johnson Space Center?

The writer remembers visiting German friends from the early astronautical congresses who were by the 1960s at Huntsville's Redstone Arsenal, later to become NASA's Marshall Space Flight Center. They showed him the early version of the Saturn launch vehicle, a "battleship" test version, ready for exhaustive analysis on the impressive static test stand, and escorted him through the various levels of that towering structure. Even more exciting was when he sheltered behind a pillar just outside one of the engineering laboratories to watch a small, scale model of the big multiengined Saturn roar deafeningly through a test firing only a few feet away. Its performance was proudly praised by a German engineer who would later figure prominently in the launching of the spacecraft to the Moon. The uninhibited nature of the test firing reminded this author of the excitement of the 1930s when amateur tests of liquid propellant rockets by members of the German Verein fur Raumschiffahrt at Berlin's Raketenflugplatz, by members of the American Rocket Society on Staten Island, and by the GALCIT group of California Institute of Technology (figure 3.1) at Arroyo Seco, Pasadena, laid the initial groundwork for humans to escape from the Earth into space.

Meanwhile, engineers were developing the mighty machines that would carry men to the Moon, and space medical researchers were working on ways to allow humans to operate effectively in the strange environment of space and on the

FIGURE 3.1   Early rocket experiments by the GALCIT group of California Institute of Technology at the Arroyo Seco, Pasadena, in 1936 paved the way for the establishment of the Jet Propulsion Laboratory which spearheaded many of the early unmanned missions to the Moon. Left to right on this October 29, 1936, photo are: Rudolph Schott, Apollo M. O. Smith, Frank Malina, Edward S. Forman, and John W. Parsons. (*JPL*)

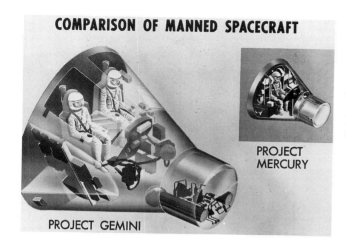

FIGURE 3.2 The first U.S. manned spacecraft were Mercury and Gemini. The two-man Gemini had 50 percent more volume than the one-man Mercury capsule. These spacecraft established the technology for space rendezvous, extended missions in LEO, and extravehicular activity in spacesuits. (*NASA*)

Moon. The space medical researchers helped to develop an impressive technology not only to protect humans from the rigors of spaceflight but also to understand many biological principles involved and how the human body reacts to changes in atmospheres, to microgravity, to reduced gravity, and to closed environments, and how humans could effectively interface with the most complex machines yet invented.

The history of the highly successful Mercury and Gemini spacecraft (figure 3.2), the precursors to Apollo, has been described often and need not be repeated here. Mercury carried lone astronauts into suborbital and later orbital flight starting in 1961. Beginning in March 1965, each Gemini spacecraft carried two astronauts into orbital flight. Suffice to say that these and concurrent Soviet missions by humans into suborbital and then orbital flight and the gaining of experience in extra vehicular activities of space walks formed the firm basis for the greatest and most challenging human adventure of all time: the first expedition to another world.

Major technical and management challenges had to be defined and overcome. Space exploration provided an unprecedented stimulation to education and to the general economy and a reward of economic multipliers that led to an era of prosperity that, unfortunately, was ended by the tremendous and seemingly futile expense of the war in South-East Asia. Fortunately, the Moon exploration program had sufficient momentum to continue for a while before the military demands on resources could eat like a malevolent cancer into worthwhile civilian projects: the exploration of new worlds, and the expansion of humankind's horizons.

• AUTOMATED PIONEERS

In July 1964, just one month before the massive U.S. involvement started in Vietnam following the Tonkin Gulf Resolution, an unmanned Ranger VII radioed back to Earth a series of increasingly close-up pictures of a lunar mare, Mare

Cognitum, until it crashed on the lunar surface. The images returned by several successful spacecraft of the Ranger program revealed impact craters in ever increasing numbers at smaller and smaller sizes (figure 3.3). Many were secondary craters produced by particles thrown out by impacts that had produced other larger craters. Ranger spacecraft returned pictures of similar detail from Mare Tranquillitatis (Ranger 8) and the interior of the large crater Alphonsus (Ranger 9). Three quite different areas of the Moon's surface were shown to be characterized by many impact craters over a wide range of sizes. The final pictures just before each spacecraft crashed showed that many of these craters were extremely small.

The next step in unmanned exploration, the Soviet Luna 9, was the first space-craft to soft land on the Moon on February 3, 1966. Its cameras confirmed that the surface was rocky and suitable for a manned landing. NASA's Surveyors also landed on the surface (figure 3.4) and returned images of their surroundings. The first of these landers, Surveyor 1, set down gently on Oceanus Procellarum in June 1966. It returned many thousands of television images, including close-ups and panoramas. Lunar soil of the same region was sampled in front of the television eye of spacecraft Surveyor 3. The next spacecraft to land, Surveyor 5, determined that the mare surface contained basalt. The final spacecraft of the series, Surveyor 7, landed close to Tycho (figure 3.5) in the lunar highlands, and found the surface there littered with rocks.

The next important step was the launch of the first U.S. orbiting spacecraft (figure 3.6a) in August 1966, which provided detailed, high-resolution photographs of much of the Moon's surface, including parts of the far side. Images returned by five Lunar Orbiters were assembled to provide the first detailed photo-map of the Moon. These spacecraft also made valuable contributions to initial maps of the Moon's gravity field. Anomalies in this field were interpreted as caused by concentrations of heavy materials (mascons) located beneath some of the circular maria.

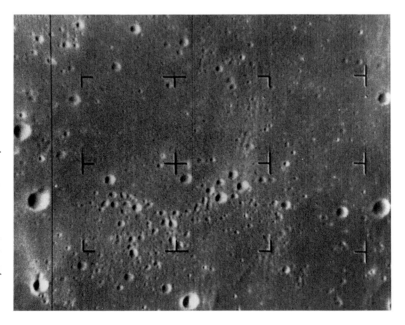

FIGURE 3.3 One of 4,316 images returned by the Ranger VII spacecraft on its way down to the surface of the Moon on July 31, 1964. The smooth-looking appearance of the mare surface was found to be pitted with many craters of ever-diminishing sizes. (*NASA*)

FIGURE 3.4 A close-up of the footpad of a Surveyor spacecraft, one of a series of unmanned spacecraft to make the first U.S. soft landings on the Moon. The impression of the footpad shows that the spacecraft bounced on its landing. (*NASA*)

FIGURE 3.5 The final spacecraft of the Surveyor series landed in the highland area near to this fresh young crater Tycho to obtain information about the highland surface for comparison with the mare surfaces. The highland surface was littered with rocks scattered by the impact that blasted the crater. (*NASA*)

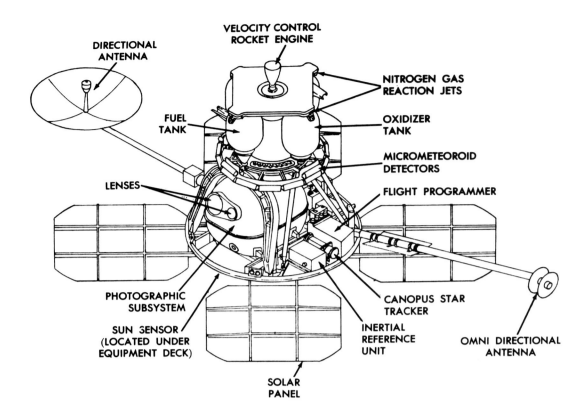

DIRECTIONAL
ANTENNA

VELOCITY CONTROL
ROCKET ENGINE

NITROGEN GAS
REACTION JETS

FUEL
TANK

OXIDIZER
TANK

MICROMETEOROID
DETECTORS

LENSES

FLIGHT PROGRAMMER

PHOTOGRAPHIC
SUBSYSTEM

CANOPUS STAR
TRACKER

SUN SENSOR
(LOCATED UNDER
EQUIPMENT DECK)

INERTIAL
REFERENCE
UNIT

OMNI DIRECTIONAL
ANTENNA

SOLAR
PANEL

LUNAR ORBITER SPACECRAFT

• THE APOLLO PROGRAM: LAUNCH VEHICLES AND SPACECRAFT

Apollo was originally conceived as a program to place a man on the Moon's surface and return him safely to Earth as a political demonstration of the superior technology of the U.S. following a spate of Soviet successes in space. Also Apollo was originally intended to make some token scientific experiments and gather samples of lunar surface materials. However, the wealth of scientific data obtained from the first manned landing stimulated a scientific onslaught on the Moon. Subsequent Apollo missions carried many additional experiments and returned increasing amounts of lunar samples. Some missions carried a roving surface vehicle so that astronauts could explore at greater distances over the lunar surface in a quest to gain a wide range of lunar samples. Also, considerable work was accomplished from Apollo's orbiting Command/Service Module during the later missions of the program.

Of nine Apollo missions three were circumlunar missions, which did not land, while six landed on the surface. Twelve astronauts walked on the surface, three rover vehicles carried astronauts up to a radius of almost eight miles (13 km) from the landing sites, and 849 pounds (385 kg) of intriguing surface materials were returned to Earth's laboratories for analysis. Many hundreds of superbly detailed

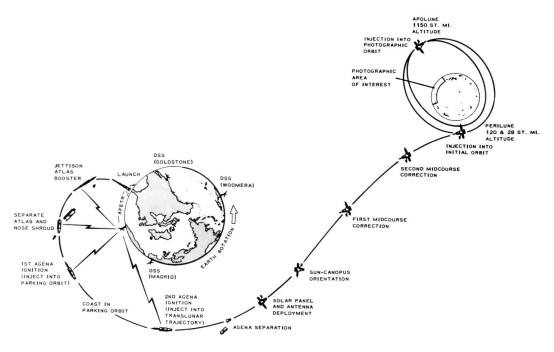

APOLUNE
1150 ST. MI.
ALTITUDE

INJECTION INTO
PHOTOGRAPHIC
ORBIT

PHOTOGRAPHIC
AREA
OF INTEREST

PERILUNE
120 & 28 ST. MI.
ALTITUDE

INJECTION INTO
INITIAL ORBIT

SECOND MIDCOURSE
CORRECTION

FIRST MIDCOURSE
CORRECTION

JETTISON
ATLAS
BOOSTER

LAUNCH

DSS
(GOLDSTONE)

DSS
(WOOMERA)

SEPARATE
ATLAS AND
NOSE SHROUD

EARTH ROTATION

1ST AGENA
IGNITION
(INJECT INTO
PARKING ORBIT)

DSS
(MADRID)

SUN-CANOPUS
ORIENTATION

COAST IN
PARKING ORBIT

2ND AGENA
IGNITION
(INJECT INTO
TRANSLUNAR
TRAJECTORY)

SOLAR PANEL
AND ANTENNA
DEPLOYMENT

AGENA SEPARATION

B. TYPICAL FLIGHT PROFILE

C. TYPICAL ORBITER IMAGE

FIGURE 3.6 (a) Drawing of the Lunar Orbiter spacecraft. Its mission was to orbit the Moon and obtain detailed photographs of most of the lunar surface in readiness for the manned landings by Apollo astronauts. (b) Several spacecraft were each launched by an Atlas Agena-D booster, and took about 90 hours to reach the Moon along the path shown in this drawing. (c) A typical Lunar Orbiter image showing the far side crater Tsiolkovsky which has been partially filled with lava. The picture is made up of a series of strips of images returned from the orbiting spacecraft. The camera system differed from later imaging spacecraft in that it used a photographic system processed within the spacecraft. Later spacecraft used purely electronic imaging systems. (*NASA photos*)

ASSAULT ON THE MOON

photographs were taken on the surface and from the orbiting Command/Service Module.

Initial U.S. planning for a launch vehicle capable of supporting a manned mission to the Moon began in 1957. By the following year studies had concluded that a clustered booster capable of developing 1.5 million pounds (0.68 million kg) of thrust was feasible. Rocketdyne, a division of North American Rockwell Corporation (now Rockwell International), developed a 200,000 pound (90,700 kg) thrust version of the H-1 liquid-propellant rocket engine used in the Thor and Jupiter intermediate range ballistic missiles (IRBM). In October 1958 the U.S. Army started to develop at Redstone Arsenal, Huntsville, a high-performance booster for advanced space missions. This was initially known as the Juno V and ultimately became Saturn. The booster was turned over to NASA in 1959 and development continued when the rocket and missile capabilities and facilities of Redstone Arsenal became NASA's Marshall Space Flight Center, Huntsville.

In July 1960, NASA proposed a manned flight program with the goals of Earth-orbital flight followed by a circumlunar mission by a three-man crew. In that same year McDonnell Douglas Corporation was selected to build a second stage of the Moon rocket designated the Saturn S-IVB (figure 3.7), and Rocketdyne was selected to develop the liquid oxygen/liquid hydrogen high performance upper stage engine (figure 3.8).

FIGURE 3.7   The McDonnell Douglas–built Saturn S-IVB high performance upper stage accelerates the Apollo Command/Service module to sufficient velocity for it to reach the Moon. (*McDonnell Douglas*)

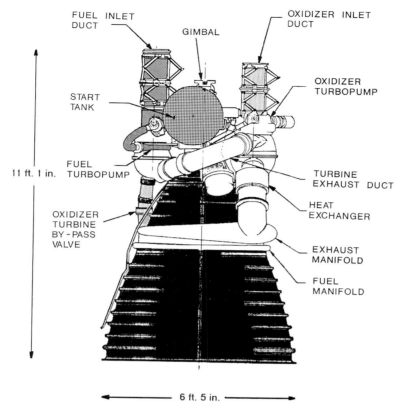

FUEL INLET DUCT
GIMBAL
OXIDIZER INLET DUCT
START TANK
OXIDIZER TURBOPUMP
FUEL TURBOPUMP
11 ft. 1 in.
OXIDIZER TURBINE BY-PASS VALVE
TURBINE EXHAUST DUCT
HEAT EXCHANGER
EXHAUST MANIFOLD
FUEL MANIFOLD
6 ft. 5 in.

FIGURE 3.8 The Rocketdyne J-2 engine burned liquid oxygen and liquid hydrogen to develop a specific impulse of 424 seconds and a thrust of 100 tons. (*NASA*)

The scene was now set for a political announcement. The space enthusiasts had done all they could to promote the need for a flight to the Moon, the beginning of human expansion into space. Now the time was opportune for political leadership to state a national goal. An historic speech to the U.S. Congress by President John F. Kennedy on May 25, 1961, established the national goal: "Now is the time," he said, "for this nation to take a clearly leading role in space achievement, which in many ways may hold the key to our future on Earth. . . . this is not merely a race. Space is open to us now; and our eagerness to share its meaning is not governed by the efforts of others. We go into space because whatever mankind must undertake, free men must fully share."

How appropriate are these words today when the nation is vacillating about its future and has so few national positive goals and so many negative ones.

President Kennedy stated, "I believe that this nation should commit itself to achieving the goal, before this decade is out, of landing a man on the Moon and returning him safely to Earth. No single space project in this period will be more impressive to mankind, or more important in the long-range exploration of space; and none will be so difficult or expensive to accomplish."

He continued; "Let it be clear . . . that I am asking the Congress and the Country to accept a firm commitment to a new course of action, a course which will last for

FIGURE 3.9 Rockwell International developed the Command/Service Module which orbited the Moon while the Lunar Module went down to the surface. The Command/Service Module brought the astronauts back to Earth where the Command Module alone reentered for a landing in the ocean. (*NASA*)

many years and carry very heavy costs . . . If we are to go only halfway, or reduce our sights in the face of difficulty, in my judgement it would be better not to go at all.''

In August of that same year the Massachusetts Institute of Technology was selected to develop the guidance and navigation system for the Moon project, and a few months later North American Rockwell was selected by NASA to be the principal contractor to develop the Command and Service Modules (figure 3.9) for the program. In the following months contracts were awarded throughout industry for the various hardware components needed to send a mission to the Moon. Chrysler Corporation was selected to make a two-stage Saturn IB booster (figure

3.10) to be used for Earth-orbital test flights. The Boeing Company was chosen to produce the first stage of the mighty Saturn V, and North American Rockwell to produce the second stage. The second stage of Chrysler's Saturn IB was to be used as the third stage of the Saturn V.

The Saturn V launch vehicle is shown in figure 3.11(a). A better idea of its size can be gained from the line drawing of 3.11(b), in which it is compared with the German V-2 (the first big rocket), other launch vehicles, and the Statue of Liberty. The three-stage booster was 281 feet (85.3 m) tall, and the spacecraft assemblage

FIGURE 3.10   The Saturn I was used to carry into orbit the Apollo boilerplate test spacecraft for qualification testing. Later it was used for other manned space missions. (*NASA*)

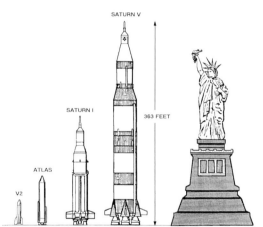

FIGURE 3.11 (*l*) The Saturn V stands alone in its launch configuration at Kennedy Space Center, Florida, ready to carry the first humans to another world. (*NASA*) (*r*) This drawing compares the sizes of the German V-2, the Atlas, Saturn I and Saturn V with the Statue of Liberty. It graphically shows the awesome technological developments from the V-2 of 1945 to the Saturn of 1969.

added another 82 feet (25 m) for a total height of 363 feet (110.6 m). The first stage was 33 feet (10 m) in diameter and was powered by five engines each of which produced 1.5 million pounds (680,000 kg) of thrust (figure 3.12) for a total thrust of 3750 tons. The second stage, of the same diameter, had five smaller engines that together produced 1.15 million pounds (522,000 kg) of thrust. The third stage had a single high performance engine that produced 230,000 pounds (104,000 kg) of thrust from liquid oxygen and liquid hydrogen propellants.

Of several possible scenarios for a voyage to the Moon, the one known as lunar orbital rendezvous was selected by NASA in July 1962. A direct flight was made from Earth's surface to an orbit around the Moon. From the lunar orbit one spacecraft would descend to the lunar surface and, after exploration, part of it

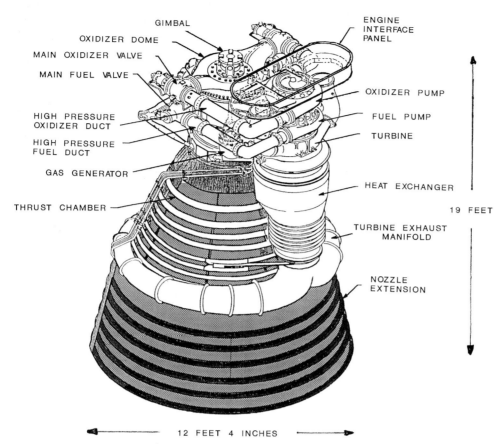

GIMBAL

OXIDIZER DOME

MAIN OXIDIZER VALVE

MAIN FUEL VALVE

HIGH PRESSURE
OXIDIZER DUCT

HIGH PRESSURE
FUEL DUCT

GAS GENERATOR

THRUST CHAMBER

ENGINE
INTERFACE
PANEL

OXIDIZER PUMP

FUEL PUMP

TURBINE

HEAT EXCHANGER

TURBINE EXHAUST
MANIFOLD

NOZZLE
EXTENSION

19 FEET

12 FEET 4 INCHES

FIGURE 3.12 The F-1 rocket engine represented an almost unbelievable development from pre–World War II rocket engines. Standing 19 feet tall it developed an incredible thrust of 680 tons, with a specific impulse of more than 260 seconds. Five of these engines powered the first stage of the Moon rocket, consuming more than 12.6 tons of liquid oxygen and RP-1 fuel every second. (*NASA*)

would return to lunar orbit to rendezvous with a spacecraft module left circling the Moon. This module would then return to Earth. The scenario called for two major spacecraft: a Command/Service Module and a Lunar Module. The Command/Service module carried the Lunar Module from Earth to lunar orbit. It provided propulsion to return from lunar orbit to the Earth. The Command Module separated from the Service Module for a reentry into Earth's atmosphere. The two-man Lunar Module carried its crew to a landing on the Moon and returned them later to lunar orbit.

The Lunar Module (LM) was a two-stage vehicle. The bottom stage contained the descent engine and a four-legged landing gear and became a launch pad for the upper stage when that stage, which contained the crew compartment, had to return the two astronauts to lunar orbit (figure 3.13). The upper stage also contained controls, life support system, and guidance and control systems, and a rocket engine for the ascent back to lunar orbit. Grumman Aerospace was selected by NASA to design and build this vehicle (figure 3.14).

FIGURE 3.13 The Lunar Module, here shown on display at the Smithsonian Air and Space Museum, was designed and built by Grumman Corporation. Other modules were carried into lunar Orbit and attached to the Command/Service Module from which they separated and descended to the Moon's surface. (*Grumman Corporation*)

North American Rockwell was selected by NASA to design and build the Command/Service Module. The Command Module was the module that returns to Earth. It contained the crew's living compartment for the voyage to and from the Moon and all controls needed for navigating there and back. The cone-shaped spacecraft was 12.8 feet (3.9 m) in diameter at its base, and it stood 12 feet (3.66 m) high. It consisted of two shells: a pressurized inner crew compartment and an outer heat shield for use during direct reentry at 25,000 miles per hour (40,230 kph) into the Earth's atmosphere on return from the Moon. In February 1966 the first full-system Apollo Command Module was successfully launched by a Saturn IB and the heat shield was proved out on reentry.

Mounted on top of the Command Module at launch was a launch escape tower equipped with rocket motors that could lift the Command Module away from the Saturn booster should there be an emergency need to do so during the launch liftoff or shortly thereafter. The escape system would have lifted the Command Module sufficiently high for parachutes to bring it and its human crew safely down to Earth's surface.

Attached behind the Command Module was a Service Module (figure 3.15), a cylinder 12.8 feet (3.9 m) in diameter and 22 feet (6.7 m) long. It contained the electrical power system for the spacecraft, oxygen and hydrogen tanks, and a primary propulsion system for maneuvers, entry into lunar orbit, and escape from

FIGURE 3.14 The lower part of the Lunar Module became the launch platform for the crew-carrying upper part. One of the most spectacular aspects of the lunar missions was the live transmission from a camera mounted on the Lunar Rover. The remotely controlled television system designed by RCA relayed a color telecast so that home TV viewers were able to watch the astronauts lift off from the Moon on their way back to Earth. (*RCA*)

FIGURE 3.15 The Command/Service Module, built by North American Rockwell (now Rockwell International), allowed astronauts to exit the Command Module and retrieve film cassettes and other science data records from the Service Module during the homeward voyage deep in cislunar space. (*North American Rockwell Space Division*)

lunar orbit. The Command Module separated from the Service Module for entering Earth's atmosphere on return from the Moon.

A Lunar Module Adapter connected the Command and Service Modules to the third stage of the Saturn V launch vehicle and served as a protective covering for the Lunar Module during launch.

The first test flight of Apollo/Saturn V, on November 9, 1967, was successful. Next a Saturn IB carried a Lunar Module on a successful test flight in January 1968, and in April of that year an unmanned Saturn V demonstrated that the launch vehicle and the various spacecraft modules were ready for flight to the Moon.

Just less than a year after the success of the unmanned flight test of a Command Module, a tragedy occurred when a Command Module undergoing a ground test with three astronauts at the controls developed an internal fire because of electrical arcing in the oxygen internal atmosphere. The three astronauts who were to have been the first to go to the Moon—Virgil I. Grissom, Edward H. White II, and Roger B. Chaffee—could not be rescued. They died in the fire and became the first Americans to give their lives in the epoch-making thrust of humankind into space and its escape from the planet on which the species evolved.

Apollo 7, the first manned flight of the complete system orbited the Earth in October 1968 and demonstrated we were ready for the first humans to travel through space to explore another world.

• VOYAGE AROUND THE MOON

Apollo 8 lifted from the Kennedy Space Center in December 1968 for the first expedition to the Moon. It did not carry a Lunar Module for a landing since this was intended as a circumnavigation mission only. Its Command/Service spacecraft carried astronauts Frank Borman, James A. Lovell, Jr., and William A. Anders 10 times around the Moon before returning them to Earth. In orbit the astronauts enthralled millions of people during the Christmas season of that year with their comments, reading from Genesis, and live television views of that other world of the Moon; its stark mountains, ubiquitous craters, and vast wrinkled plains presented evidence left undisturbed since a violent creative event aeons ago.

The first complete mission with the Lunar Module as well as the Command/Service Module was Apollo 9, launched in March 1969 into Earth orbit. During the orbital flight the separation of the Lunar Module from the Command/Service Module, its reorientation, docking and transfer of astronauts to and from it, were all demonstrated successfully.

One more critical test was needed before a landing on the Moon would be attempted. Apollo 10 was launched in May 1969 as a lunar mission in which everything would be tried except an actual landing on the lunar surface. The Lunar Module separated successfully from the Command/Service Module and entered an elliptical orbit carrying it down toward the Moon. It descended to an altitude of 9.7 miles (15.6 km) above the site in Mare Tranquillitatis where the first landing would be made. While astronaut John W. Young remained orbiting the Moon in the Command Module, astronauts Thomas P. Stafford and Eugene A. Cernan were able to inspect the site at close range as they traveled across the lunar surface. Apollo 10 made 31 revolutions around the Moon before returning to Earth.

• FIRST EXPLORATIONS ON THE SURFACE OF ANOTHER WORLD

The launch of Apollo 11, the first spacecraft to attempt to land humans on the Moon, took place July 16, 1969. On arrival at the Moon the Command/Service Module and attached Lunar Module were placed in an elliptical orbit inclined 1.25 degrees to the lunar equatorial plane. The rocket engine of the Service Module was ignited to change the module's path to an almost circular orbit with a mean height of approximately 37.3 miles (60 km) above the lunar surface and with a period of two hours. Many photographs were taken of the lunar surface, to be used for lunar mapping and geodetic studies needed for planning and executing later missions.

Next the Lunar Module with its two astronauts, Neil A. Armstrong and Edwin E. Aldrin, Jr., separated from the Command/Service Module leaving Michael Collins to continue orbiting alone. The descent engine of the Lunar Module was ignited for a short burn, and the spacecraft moved down toward the lunar surface. Final ignition of the descent engine slowed the module further and the engine continued burning until touchdown on the lunar surface. The first manned landing was highly successful with touchdown occurring exactly 102 hours, 45 minutes, and 40 seconds (5.011 days) after launching from Kennedy Space Center, Florida.

On July 20, 1969 the Lunar Module landed in the southwestern part of Mare Tranquillitatis in a relatively flat area across which were bright rays from a sharp-rimmed blocky crater and some ejecta from the large crater Theophilus. Humanity's first temporary base, *Tranquility Base,* had been established on another world when Neil A. Armstrong descended from the ladder of Apollo 11's Lunar Module. This first step onto another world and the subsequent expeditions by Armstrong and Aldrin on the Moon's surface were broadcast worldwide on television and radio. The event was watched live by more people than any other event in Earth's history.

The first scientific objective of Apollo 11 was to collect as quickly as possible a sample of the lunar surface for return to Earth. This was a contingency sample in case the mission had to be aborted and the astronauts had to leave the Moon soon after their arrival there. Further samples were then gathered and instruments placed on the lunar surface. Photographs were taken of the surface with a 70–mm camera and with the lunar stereoscopic closeup camera. A period of 2 hours and 40 minutes was allocated for activity on the Moon. All planned tasks of the surface mission were completed satisfactorily, and 44 pounds (20 kg) of lunar material were collected.

On completion of their exploration the astronauts returned to the Lunar Module. Its ascent engine was ignited and the astronauts started their long journey back to their home planet. A series of maneuvers began that resulted in the Lunar Module docking with the Command/Service Module in lunar orbit. Armstrong and Aldrin transferred into the Command Module taking with them the lunar samples and

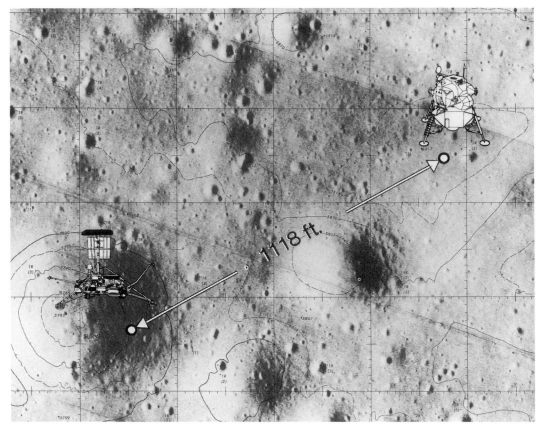

1118 ft.

FIGURE 3.16 Planning for Apollo 12 intended a landing just over 1,100 feet (340 m) from Surveyor III in Oceanus Procellarum. (*NASA*)

photos. Next, the Lunar Module was jettisoned and the engine of the Service Module was ignited to break the spacecraft from lunar orbit and head it back to Earth.

Several days later, in a final maneuver, the Service Module separated from the Command Module. The Command Module with its three astronauts and science payload entered Earth's atmosphere at 26,723 miles per hour (43,000 kph) to decelerate by aerobraking and parachutes and to splash down in the Pacific Ocean 195 hours, 18 minutes, and 35 seconds (13.446 days) after launch. The age-old dream of humans journeying to the Moon had at last become a reality.

In an introduction to the NASA report on the preliminary scientific results from this first landing on another world, Robert R. Gilruth and George M. Low pointed out that the mission to the Moon was extremely complex from all points of view. It required extensive planning to enormous detail. It required reliable hardware of the highest available technology. It required advanced software for a variety of digital computers, and it demanded effective and efficient operations. "In parallel with the emphasis on engineering problems and their solution, the scientific part of the Apollo 11 mission was planned and executed with great care," they wrote. "The success of the Apollo 11 mission was solidly based on excellent technology, sound decisions, and a test program that was carefully planned and executed."

"To this foundation was added the skill and bravery of the astronauts, backed up by a fully trained and highly motivated ground team."

The next Apollo mission, Apollo 12, was ready to go within four months. The landing site was in Oceanus Procellarum (Ocean of Storms) at 2.94 degrees south latitude and 23.45 degrees west longitude; about 800 miles (1300 km) west of *Tranquility Base*. An added bonus for choosing this site was that Surveyor 3 landed on the inner edge of a crater about 1100 feet (335 m) from the Apollo 12 aiming point (figure 3.16). Retrieval of components from that unmanned spacecraft would reveal how they had survived almost three years in the lunar environment.

Liftoff from Kennedy Space Center took place on November 14, 1969, marred by a lightning strike hitting the spacecraft as it rose into the clouds. The astronauts recovered the launch sequence and were able to continue with the mission, demonstrating the advantage of humans over machines. Astronauts Charles Conrad, Jr. and Alan L. Bean descended to the lunar surface while Richard F. Gordon remained in the Command Module. The astronauts maneuvered their spacecraft to touch down about 600 feet (183 m) from Surveyor (figure 3.17). They could have landed

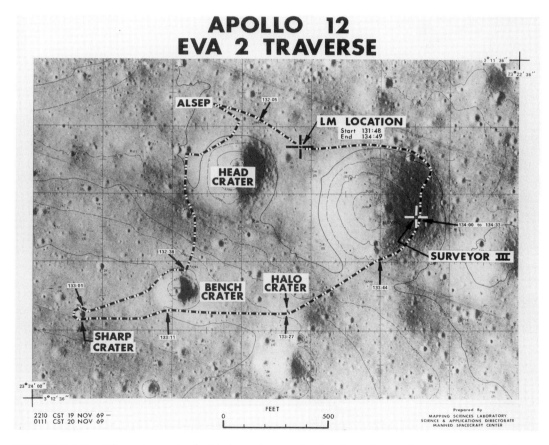

FIGURE 3.17 The astronauts decided to change the landing site as they approached closer, they maneuvered around a crater and touched down with pinpoint accuracy only 600 feet (183 m) from the Surveyor, demonstrating superb piloting skill in landing the Lunar Module spacecraft. Also shown on this illustration are the paths of the traverses made by the astronauts over the lunar surface. (*NASA*)

closer but they did not want to blow lunar dust over the unmanned spacecraft. Their landing demonstrated pinpoint precision, actually maneuvering around a crater and landing away from the preselected site, which on close inspection during the descent was considered by the astronauts as too rough.

Two explorations were made on the surface each of which lasted for four hours outside the Lunar Module (see figure 3.17). All the planned traverses over the surface were successful. Surface samples were collected, and instruments set up on the surface to continue operating after the astronauts returned to Earth.

One problem of operating on the Moon became apparent: The insidious nature of the ubiquitous lunar dust that infiltrated everywhere. The astronauts decided to remain in their suits between each extravehicular activity to safeguard against the dust getting into vital parts of the spacesuits. When they visited the Surveyor spacecraft (figure 3.18) they discovered that the Lunar Module's engines had kicked

FIGURE 3.18 The Surveyor III spacecraft was photographed by the Apollo 12 crew. They later removed wiring and the TV camera for inspection back on Earth. (*NASA*)

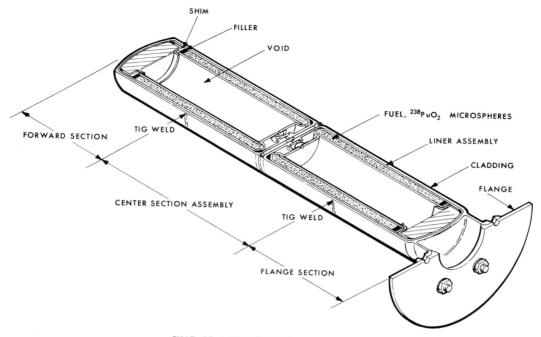

SNAP-27 RADIOISOTOPIC FUEL CAPSULE

FIGURE 3.19 Cut-away view of the SNAP-27 fuel capsule that was fueled and welded at Mound Laboratory of Monsanto Research Corporation for use with Apollo. Heat given off from the decay of plutonium-239 dioxide was converted by the thermoelectric generator into electricity to power instruments on the Moon. The capsule was about 16 inches (40 cm) long. (*Monsanto Company*)

up sufficient dust for the pristine white of the Surveyor to be coated with dust, which gave it a tan or brown color. The astronauts cut samples of aluminum tubing from the spacecraft together with some electrical cables, and they unbolted the camera. These three types of spacecraft equipment were brought back to Earth to be examined by engineers and scientists. In orbit astronaut Gordon took many photos of the lunar surface to provide detailed information of landing sites proposed for subsequent missions.

There were several firsts on this mission. The Lunar Module carried a radioisotope thermoelectric generator (SNAP-27) which was placed on the Moon to provide a year's supply of electrical power for instruments left there by the astronauts; the first use of a nuclear power device on another world.

The fuel used was plutonium-238, fully oxidized to plutonium dioxide, which is chemically and biologically inert. The rugged capsule housing the fuel (figure 3.19) was designed to survive reentry heating and to contain the fuel in the unlikely event of an abortive launching of the Saturn V and the spacecraft from Kennedy Space Center. An additional protective cask contained the fuel capsule. The astronauts removed the fuel unit from the cask when it was safely on the Moon (figure 3.20). An astronaut then placed the capsule into the generator in which lead telluride thermoelectric elements converted into electrical energy the heat produced by alpha particles arising from the radioactive decay of the plutonium dioxide.

For the first time the astronauts' life support system backpacks were recharged so they could be used again on a second traverse of the surface. These backpacks

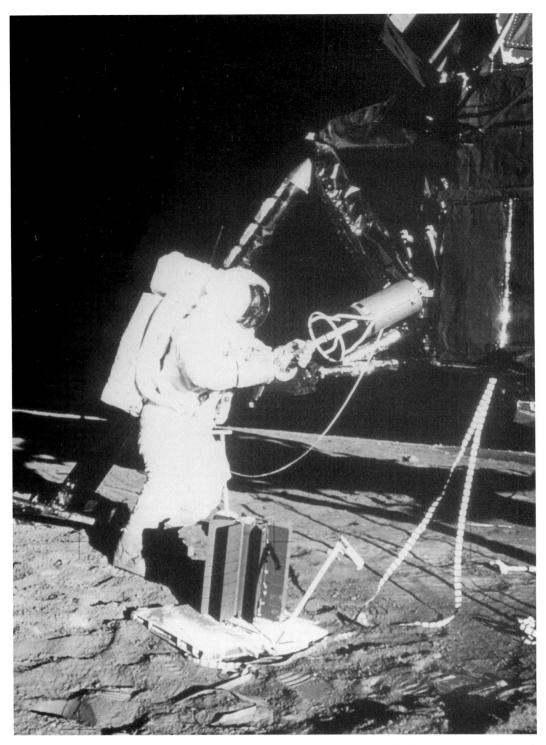

FIGURE 3.20 Astronaut Bean extracts a plutonium–238 fuel capsule from its protective cask on the side of Apollo 12's Lunar Module to insert it into the first nuclear-powered thermoelectric generator to supply electricity for experiments on the Moon. (*NASA*)

supplied each astronaut with oxygen, water (used to cool the suit), and electrical power. In the time between the walks on the lunar surface while the astronauts were in the Lunar Module, the water and oxygen tanks were refilled from the Lunar Module supplies, and the backpack silver-zinc batteries were replaced. The batteries supplied electricity for the oxygen-circulating fan, the water pump, and radio communications. Also the cartridges of lithium hydroxide used to absorb carbon dioxide from the internal atmosphere of the spacesuits, were changed.

An interesting discovery was made later on Earth at the Lunar Receiving Laboratory when scientists examined the returned parts of Surveyor. Microbial analysis revealed that a microorganism—streptococcus mitis—had survived for 950 days in the lunar environment within the polyurethane foam circuit board insulation of the camera.

The microorganism, which is a common respiratory organism, was incubated and soon started to reproduce actively in the laboratory. It was thought that microorganisms were inadvertently deposited within the camera during one of the times the camera shroud had been removed for repairs before the launching of the Surveyor. The microorganism had survived because it reverted to a lyophilized state equivalent to freeze drying.

Also examination of the samples brought to the Lunar Receiving Laboratory revealed that one rock had 20 times as much uranium, thorium, and potassium as any other sample so far examined. The three-ounce (83–gm) rock also appeared to be the oldest rock so far identified; 4.6 billion years old. Other rocks examined to that time were between 3.3 and 3.7 billion years old. The unusual specimen was of the age established for meteorites, which was thought to be when the Solar System formed. The rock was believed to have originated from the impact that produced the fresh crater Copernicus and spread material from deep within the lunar crust.

The success of Apollo 12 resulted in a congratulatory letter from N. Podgorny, the President of the Presidium of the USSR Supreme Soviet, to U.S. President Richard M. Nixon. ". . . congratulations in connection with the successful completion of the flight of the spacecraft Apollo 12, which has become a new step in the exploration of the moon by man."

Also Konstantin Feoktistov, a Soviet Cosmonaut, in an interview to *Izvestia*, stated that he did not share the opinion, expressed in much of the U.S. press, that the Apollo-11 flight yielded practically nothing new from a scientific angle. "This assertion is unfair," he said. "The astronauts brought back from the moon samples of lunar rock, and their initial study yielded interesting scientific results. Commenting on the Apollo 12 flight he added, "In my opinion, the main importance of the Apollo-12 flight is the study of the moon in various places. Apollo-12 was the next step after Apollo-11, and it was made quite successfully. All of us congratulate our American colleagues on this success." He also stated that: "The results are excellent. Making a pin-point landing on the moon at the present stage of advancement of cosmonautics is truly a great technical accomplishment. Credit for this, naturally, belongs not only to Charles Conrad and Alan Bean, who showed very high standards of spacecraft flying, but also to the big group of American specialists who built the lunar module."

The Vice-President of the USSR Academy of Sciences, Academician Alexander Vinogradov also said: "This is a great success of modern science and technology. It has opened new possibilities for science in probing the universe. . . . An analysis of the specimens of Moonrock has revealed, notably, that they crystallized from molten mass that erupted from the lunar interior on the surface. Hence, we can draw the conclusion that the moon has its own mantle, [and] a lunar crust, whereas most of the models of the moon, made before the Apollo flights, presented our natural satellite as a solid homogeneous mass, with no pronounced crust."

The next mission, Apollo 13, was intended to land in the Frau Mauro area of the Moon. An unfortunate explosion within the liquid oxygen tank of the Service Module on the way to the Moon aborted the mission. By use of tremendous ingenuity on the part of the three astronauts—James A. Lovell, John L. Swigert, Jr. and Fred W. Haise, Jr.—and with enormous support from Mission Control at Houston and scientists and engineers of many disciplines—the crew was able to urge the crippled spacecraft around the Moon and bring the Command Module to a safe return to Earth.

The author, covering this mission for a major newspaper, wrote: "As a scientific mission Apollo 13 was almost, but not quite, a total failure. As a demonstration of brave men's capabilities to face and meet dangers of space operations it was an unqualified success." Indeed, the successful return of Apollo 13 demonstrated the superiority of manned over unmanned missions when troubles arise as they inevitably seem to do in spite of the most careful planning. While there have been several spectacular recoveries of ailing automated spacecraft that went on to complete their missions, none has faced a catastrophic failure similar to that of Apollo 13. An unmanned spacecraft suffering the failure of Apollo 13 would have been lost. The astronauts reconfigured parts of the spacecraft with great ingenuity to keep it operating and ensure a safe return. They were even able to complete some science by photographing the lunar surface as the crippled spacecraft went around the Moon.

The problems that caused the Apollo 13 mishap were identified and corrected, and the spacecraft for the next mission was prepared for launch. The landing site chosen for Apollo 14 was that intended for Apollo 13. It was in the hilly uplands north of the crater Frau Mauro (figure 3.21) about 110 miles (180 km) east of the landing site for Apollo 12. The area was expected to provide new information about the age of the Moon. The Frau Mauro formation on which the landing took place is a blanket of debris that may have originated from 100 miles (160 km) or so beneath the lunar surface. This material was excavated when a large impacting body produced the Mare Imbrium basin and spread ejecta far over the surrounding surface. The blanket of material in the Frau Mauro formation probably came from Imbrium as ballistically ejected material and an outward surge of gas and debris.

The area selected for the landing of Apollo 14 is characterized by ridges a few hundred feet high radiating from the Imbrium basin. The ridges are separated by undulating valleys. The blanket of material originally ejected at the formation of Imbrium is today buried beneath a layer of younger rubble and lunar soil churned up by subsequent impacts. Near the chosen landing point one of these more recent impacts pierced the covering layer of debris to produce a 1000–foot (300–m) diameter fresh crater identified as cone crater. Huge blocks of original Imbrium material were scattered around the rim of this fresh crater.

Besides placing science equipment to remain on the surface, and gathering of surface samples, the mission produced data about the Moon's structure to depths

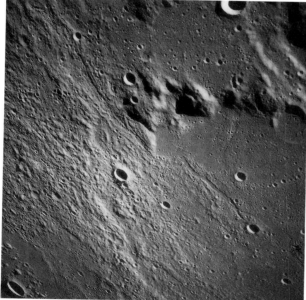

FIGURE 3.21 The area selected for landing of Apollo 14 was to explore beyond the lunar maria into the blanket of debris which was flung across the Moon by the impact which created the Imbrium Basin. (t) A general view of the area looking north across the ancient ruined craters Frau Mauro (top-most of the three), Parry, and Bon-pland. (b) Vertical view of the northern part of Frau Mauro showing the smooth, almost rectangular area near the center where a spacecraft might land safely. North is at the top. (*NASA photos*)

far beyond the earlier missions. This was accomplished by allowing the 30,000 pound (13,600 kg) Saturn V third stage to crash onto the Moon at 5800 miles per hour (9300 kph) at a location 100 miles (160 km) farther from the Apollo 12 seismic instrument than had the spent Saturn stage of the ill-fated Apollo 13. The increased range permitted the seismic waves to travel deeper into the lunar crust. After its two astronauts had transferred to the Command Module for return to Earth, the Lunar Ascent Module also was allowed to impact on the Moon and create seismic waves.

On the Moon the crew set off 21 small surface explosions to generate seismic waves. They also set up a mortar that fired four grenades after the astronauts left the surface. Fired by command from Earth, the grenades produced explosions on the lunar surfaces at distances of 500, 1000, 3000, and 5000 feet (150, 300, 915, and 1525 m) from the seismic detectors. The vibrations generated by the explosions were picked up by geophones and the data were transmitted to Earth.

For the Apollo 14 exploration, Grumman Aerospace developed a two-wheeled cart that the astronauts wheeled across the surface (figure 3.22), to help carry science equipment and samples.

FIGURE 3.22 The exploration at Frau Mauro site was aided by the astronauts having a two-wheeled cart developed by Grumman Corporation. This illustration by Craig Kavafes shows the astronauts beginning a walk to set up instruments and gather samples using the cart. (*Grumman Corporation*)

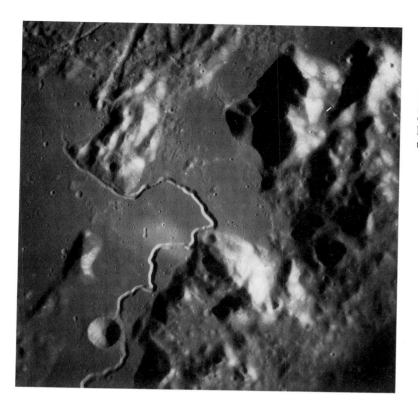

FIGURE 3.23 Apollo 15 landed close to the Hadley Rille, a sinuous canyon some 60 miles (100 km) long and in parts 600 feet (180 m) deep. It borders the foothills of the high mountain blocks. (*NASA*)

For the next Apollo mission a northern lunar plain cut by a winding gorge at the base of some of the highest mountains on the Moon was chosen. The landing site for Apollo 15 was called Hadley-Apennine, a location about three degrees east of the center of the lunar disc as seen from Earth and about 25 degrees, or 465 miles (748 km) north of the lunar equator. All previous Apollo landings had been within 70 miles (113 km) of the equator. The mission to the Hadley-Apennine region presented a unique opportunity for the astronauts to obtain samples and make observations relating to three vexing problems about the Moon.

Actually, NASA considered five possible landing sites before picking the Hadley-Apennine area. The site chosen certainly provided some spectacular scenery as well as important geological features. The Lunar Apennine, a major mountain range, has peaks rising majestically 8000 feet (2400 m) above the flat lava flows, similar to the escarpment of the Rockies, pushing skyward from the plains of Colorado.

At the base of these mountains is a sinuous canyon (figure 3.23), the Hadley Rille, some 60 miles (100 km) in length and 600 feet (180 m) deep. The landing site for Apollo 15 was chosen close to a sharp bend in the canyon, and about three miles (5 km) north of the foothills of a 4000–feet (1200–m) high mountain block. A few miles away the highest mountain peak, Mount Hadley, rises some 10,000 feet (3000 m), surrounded by companion peaks of lesser height.

Landing at the chosen site presented some tricky piloting problems for the astronauts, James B. Irwin and David R. Scott, who had to guide the Lunar Module over the sharp, high-rising ridges of the Apennine Mountains for hundreds of miles, and then drop over the last precipitous escarpment to touch down within a mile of the rille. A normal approach path would have taken the module to within 3000 feet (1000 m) of the summit of a pass. Controllers decided this was too hazardous and increased the angle of approach from 12 to 25 degrees. This allowed a clearance of 10,000 feet (3000 m), which was accepted as providing an adequate safety margin. The steeper approach path also allowed a landing at a time when the Sun was higher in the lunar sky so that the landing site would not be obscured by the long shadows stretching from the Apennine Mountains across the Imbrium plain of Palus Putredinis.

The site gave access to a treasure of geology for the astronauts to explore. First, they were able to collect materials from the base of the Apennine Mountains, which generally rise more than 8000 feet (2400 m) from the mare surface near to the landing point. Such samples were expected to contain a mix of the old lunar crust that existed before the Imbrium Basin was excavated and of rocks from deep within

FIGURE 3.24   The rille is believed to have been eroded by flowing lava. It was inspected by the astronauts who drove their Lunar Rover to its edge to obtain rock samples and detailed photographs. (*NASA*)

FIGURE 3.25 The inner slopes of Hadley Rille were found to have layers of basaltic lava, confirming that the plains were deposited as a series of lava flows. Outcrops show lunar bedrock where the regolith is thin at the edges of the rille. The layers vary between hard and soft rock, which has resulted in different weathering rates. At the bottom of the rille debris fallen from the sides is mixed with regolith. (*NASA*)

the Moon that were excavated by the impact. Besides extending the time-scale back beyond the 3.7 billion years average age of the samples returned by Apollo 11, the rocks were expected to be significantly different in composition from the basalts obtained at the sites of Apollo 11 and Apollo 12 landings.

Second, the astronauts would journey to the Hadley rille (figure 3.24) in an attempt to obtain evidence bearing on the origin of rilles. Scientists believed that such rilles are collapsed lava tubes. Whereas on Earth such tubes do not generally last long because of weathering, on the Moon they would last relatively undisturbed for many millions of years. Lava tubes that have survived in desert regions of Earth are roughly circular tunnels leading downslope from volcanic cones. Sometimes these tubes meander and often, when the roofs collapse, they form a series of crater-shaped depressions. A good terrestrial example is in New Mexico where there are eight lava tube systems containing miles of tunnels beneath the lava fields at Baldera. Other examples are in Washington State, Idaho, and Hawaii. Confirming that Hadley Rille is a collapsed tube would be very important to future activities on the Moon. There probably would be many uncollapsed tubes also, preserved beneath the lava sheets. These would be extremely valuable assets when establishing bases on the Moon.

Visual observations and photography of the walls of the rille (figure 3.25) should confirm if there was layering, which could indicate whether the rille was a collapsed lava tube or the result of erosion by volcanic ash flows or by surface materials that flowed like fluids because of degassing from the Moon's interior. Layering was indeed found.

Third, the sampling of the fresh-looking mare and volcanic looking features at this location was expected to extend the age scale established by the Apollo 11 and Apollo 12 missions and find rocks of younger ages on the fresh-looking lava flows.

Apollo 15 was the first of what was termed the "J" Mission Series designed to explore the Moon scientifically, as contrasted with developing and testing the technology to land humans on the Moon and return them safely to Earth. Many additional items of equipment and changes to software were incorporated in the spacecraft systems. Apollo 15 was also carrying a Lunar Roving Vehicle (figure 3.26) built by Boeing to transport two astronauts and their equipment to prospective important points established before the mission, and to targets of opportunity discovered during the mission.

In June 1971 I watched the astronauts practicing using the rover at Volcano Peak in California's Mojave Desert. Because the three astronauts who would take Apollo

FIGURE 3.26   Apollo 15 carried a Lunar Rover, which allowed astronauts to make more and longer traverses to explore the surface. Such vehicles and their developments will be vital for long-term exploration from a lunar base. (*NASA*)

15 to the Moon were not scientists, but test pilots, NASA rushed to give specific training to them. The crew worked hard with lunar scientists in several exercises to try to bring maximum scientific results from the mission, especially since the Apollo program was by this time destined to end with Apollo 17.

In the exercise on the lava flows from Volcano Peak, geologists worked with the astronauts, asking them to describe over the Lunar Rover's radio communications system the geological formations they were seeing. A critical test when the astronauts arrived on the lunar surface would be if the geologists here on Earth could work through the hands and eyes of the men on the Moon. Hidden from the geologists by a ridge of lava during the tests at Volcano Peak, the astronauts began their practice. Both astronauts and scientists were provided with degraded aerial photos of the lava flow to simulate the maps they would have of the lunar surface. The exercise went on for several hours in the hot desert. But the astronauts were enthusiastic that they were riding rather than walking.

When on the Moon they could use the Rover for up to 4.7 miles (7.5 km) radially from the Lunar Module. Although the Rover could travel 23 miles (37 km) the traverses were restricted to a distance that would allow the astronauts to walk back to the Lunar Module if the Rover should develop mechanical trouble.

On the Moon later, three traverses with this rover covered a total distance of the full 23 miles (37 km) compared with the previous Apollo mission where astronauts Alan Shepard and Edgar Mitchell walked only 2.1 miles (3.38 km) from the landing site of Apollo 14.

It was intended that the Apollo 15 astronauts would test for heat flows near the edge of the Mare Imbrium and would check if there were gas emissions from the Hadley rille where gases might still be escaping from the interior at the edges of impact basins.

To accomplish this science task, the astronauts used a drill for the first time on the Moon. With it they drilled two 1–inch diameter, 10–feet deep holes into the rubble of surface material. Into each hole they placed a pair of extremely sensitive electronic thermometers (figure 3.27). The temperatures measured with these instruments were relayed to Earth by radio, and from them scientists determined the heat flow from the lunar interior and the thermal conductivity of the surface material.

The electric drill was powered by silver-zinc batteries and was supported by the astronauts holding two handles similar to a jack hammer. As the astronaut pressed the drill vertically into the surface, 22–inch sections of a hollow bore stem were added to provide a casing for the holes into which the thermometers were placed.

A third hole was drilled about seven feet deep to extract cores within six 17–inch sections from the surface layer and thus provide samples that could not be obtained by gathering them on the surface itself, as had been done on the earlier missions. The sections were loaded with the cores by a tungsten drill bit at the tip that admitted the lunar material into the sections as the drill cut through the soil and rock.

Apollo 15 also carried a new set of orbital sensors in its Command/Service Module including cameras, geochemical indicators, and a sub-satellite (figure 3.28). The orbital surveillance was improved because of the high northern latitude of the landing site, which meant that the Command/Service Module swept over a wide

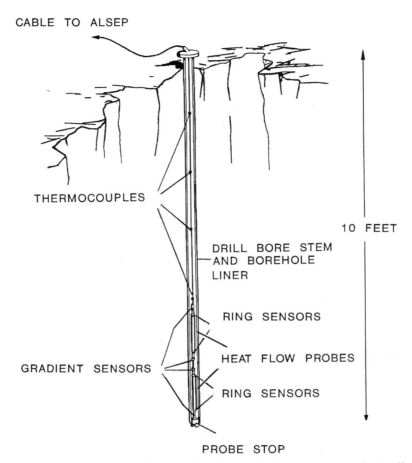

CABLE TO ALSEP

THERMOCOUPLES

10 FEET

DRILL BORE STEM
AND BOREHOLE
LINER

RING SENSORS

HEAT FLOW PROBES

GRADIENT SENSORS

RING SENSORS

PROBE STOP

FIGURE 3.27 An important experiment of the later missions starting with Apollo 15 was to measure heat flow from the interior of the Moon. This drawing, adapted from information from Arthur D. Little, Inc., shows how sensors were installed beneath the surface in a drill hole to send information back to the ALSEP central station.

swathe of the surface including a significant part of the far side of the Moon. The 80–pound (36.2 kg), hexagon-shaped sub-satellite was ejected from the Service Module early in the mission and it remained in lunar orbit for many months after the astronauts returned to Earth. When released by the firing of explosive bolts, coiled springs pushed it away from the Service Module so that it entered an orbit that carried it between 55 and 75 miles (88.5 and 120.7 km) over the lunar surface. Its particle detectors and magnetometer provided data to supplement and correlate with the surface magnetometer and particle detectors. The satellite also carried an S-band transponder so that its position could be pin-pointed for lunar gravity studies. The high inclination enabled the satellite to fly over several mascons, and this improved on the gravity data obtained by the Lunar Orbiters. The gravity measurements were also improved over those made from Lunar Orbiters because the sub-satellite was in a lower orbit.

A mountainous highland region of the Moon was selected as the exploration site for the Apollo 16 mission. It was north of the crater Descartes and a key site to complement the earlier expeditions to the lunar maria and to the uplands of Frau Mauro. Descartes, which lies in the central lunar highlands, provided samples from the material filling highland basins that was thought to be derived from two young large craters near the landing site and to be younger than the lava filling the mare basins.

Another sampling area in the region was a hilly, grooved, and furrowed terrain that was thought to be of volcanic origin. The area was expected to provide information about the interior composition of the thick highlands crust.

A Lunar Roving Vehicle was carried on this mission also, and the Service Module carried a small satellite that was ejected into lunar orbit during the mission and remained in orbit after the return of the astronauts. As with the sub-satellite launched by Apollo 15, its particle detectors and magnetometer provided data to correlate with similar instruments left on the surface. It also carried an S-band transponder so that tracking of the orbit would provide more information about the Moon's gravity anomalies.

FIGURE 3.28 Sub-satellites, launched from Service Modules into separate orbits around the Moon, were used in later Apollo flights to measure global magnetic fields and gravitational anomalies. Developments of such unmanned lunar satellites will also be needed as precursors to the establishment of the lunar outpost. (*Rockwell Space Division*)

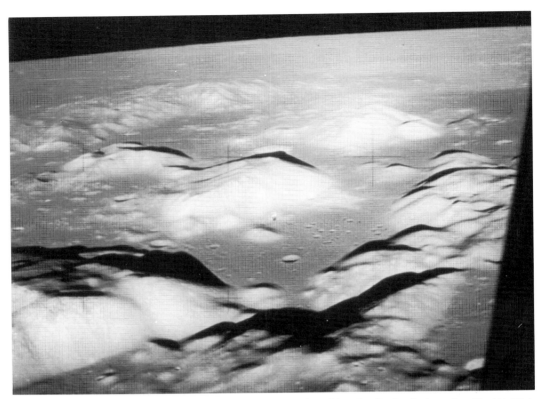

FIGURE 3.29 An important geological area was explored by the final mission, Apollo 17. This was the Taurus-Littrow Valley, a region of mountainous highlands and lowland valleys southeast of Mare Serenitatis. (a) A general view of the area taken from orbit.

The mission of the final Apollo, Apollo 17, was to explore a region of the Moon characterized by a combination of mountainous highlands and lowland valleys southeast of Mare Serenitatis (figure 3.29). This Taurus-Littrow area offered a combination of mountainous highlands and valley lowlands from which samples could be gathered. Three rounded hills dominate the area. These massifs surround a relatively flat valley, which was the landing site. In the valley, too, is a range of sculptured hills.

In company with a group of other writers and scientists, I witnessed the night launch of Apollo 17 from eight miles off the coast of Florida. Deep emotions colored the comments afterward, even from those among us who had witnessed the beginning of other missions to the Moon. Krafft Ehricke, an engineer associated with the German V-2 project at Peenemunde during World War II and at the time of Apollo a senior scientist at North American Rockwell, pointed out that human expansion into space is a natural process essential for continued development from the biosphere here on Earth.

Several of us discussed why the American public had by then turned its back on the space program in particular and science in general. Robert Heinlein, dean of science fiction writers, likened the current turning away from science as a pause rather than an end. He was supported by Marvin Minsky, Massachusetts Institute of Technology's expert on artificial intelligence, who stated that the present wilder-

FIGURE 3.29 (b) Scientist/astronaut Harrison H. Schmidt stands beside a huge split boulder ejected from an impact into the valley, with some of the surrounding mountains forming the distant backdrop, starkly etched against the black lunar sky. (*NASA photos*)

ness of human thought was like an "exploropause," but he added that scientists had to keep up their enthusiasm as they crossed the waterless desert of public opinion existing at that time.

Unfortunately the drought in forward thinking has continued to the present time and programs such as space station Freedom and bases on the Moon and Mars and other major science and technology programs come under frequent attack.

Norman Mailer claimed that the Apollo program was the most heroic endeavor of the U.S. since the settlement of the West. He pointed out that the proportion of the gross national product spent on the transcontinental railroad was about the same as the proportion spent on Apollo. But he lamented that one of the most profound events of the twentieth century had been made monumentally boring and totally depressing by Washington bureaucrats. Others claimed that NASA was not entirely to blame, that the mass media failed to explain the program properly, giving more attention to unimportant factors, analogous to talking about plumbing without explaining what the pipes were carrying and why.

One of the writers stated that Soviet coverage in major newspapers of their probes, Mars 2 and 3, was much better than American media coverage of Mariner 9 which made significant discoveries about Mars. Soviet coverage was technically detailed and talked up to the reader whereas U.S. coverage of space missions was either sensational or meager and generally talked down to the reader. He stressed

TABLE 3.1 *Rationale for Site Science*

	Apollo 11	Apollo 12	Apollo 14	Apollo 15	Apollo 16	Apollo 17
Type	Mare	Mare	Hilly upland	Mountain front/ rille/ mare	Highland hills and plains	Highland massifs and dark mantle
Process	Basin filling	Basin filling	Ejector blanket formation	Mountain scarp Basin filling Rille formation	Volcanic construction Highland basin filling	Lowland filling Volcanic mantle
Material	Basaltic lava	Basaltic lava	Deep-seated crustal material	Deeper-seated crustal material Basaltic lava	Volcanic highland material	Crustal material Volcanic deposits
Age	Older mare filling	Younger mare filling	Early history of Moon Pre-mare material Imbrium basin formation	Composition & age of Apennine front material Rille origin & age Age of Imbrium mare fill	Composition & age of highland construction & modification Composition & age of Cayley formation	Composition & age of highland massifs and lowland filling Dark mantle composition & age Rock landslide

that a major threat to the U.S. was disenfranchisement through ignorance. People had nowhere to go to find real facts on anything technical. A communications crisis was deluging the people with colorful trivia.

Roger Caras, vice president of the company that produced the motion picture *2001—A Space Odyssey,* stated that the American public had turned away from many other important things as well as the space program. He suggested that humankind may be entering a very critical stage of human experience. "We are so over communicated that we cannot maintain interest in anything beyond the novelty stage," he said. How true many of these comments continue to apply even today some 20 years later.

A comparison of the various Apollo missions and the science rationale for them is given in table 3.1. The locations of the landing sites are shown on figure 3.30.

• INSTRUMENTING THE MOON

Besides the human exploration, gathering of samples, and extensive photo surveys of lunar features from the surface and from orbit, an extremely important part of the Apollo program was the placement of Apollo Lunar Surface Experiments Packages (ALSEP) at five locations on the Moon (table 3.2) for synoptic gathering of data between Apollo missions and after the last mission.

The ALSEP was a self-contained science station placed by the astronauts and left unattended after they returned to Earth. The original service lifetime was for one year, two years for Apollo 17, but the equipment actually operated for much longer until the whole system was shut down on September 30, 1977. The objectives of

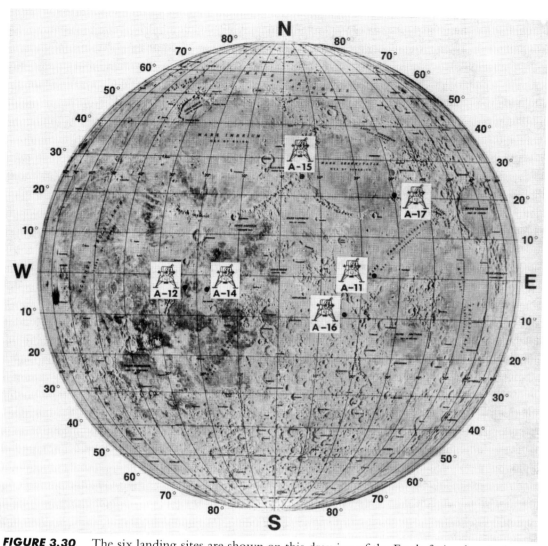

FIGURE 3.30 The six landing sites are shown on this drawing of the Earth-facing hemisphere of the Moon. ALSEP instrument groups were placed at all sites except for that of Apollo 11 in Mare Tranquillitatis. (*NASA*)

TABLE 3.2 *Locations of ALSEP*

Mission	Landing Site	Coordinates		Date
12	Oceanus Procellarum	23.5 W	0.7 S	11/19/69
14	Frau Mauro	17.5 W	3.7 S	2/05/71
15	Hadley Rille	3.7 E	26.1 N	7/31/71
16	Descartes	15.5 E	9.0 S	4/21/72
17	Taurus-Littrow	30.8 E	20.2 N	12/12/72

NOTE: In addition, a rudimentary package containing a single experiment to detect lunar seismic disturbances and a dust detector and deriving power from solar cells was deposited by Apollo 11 on Mare Tranquillitatis at 23.4 deg. E and 0.7 deg. N.

the program were fivefold. First, attempt to determine the internal structure and composition of the Moon. Second, determine the composition of the lunar atmosphere. Third, obtain new insights into the geology and geophysics of the Earth by obtaining comparative data concerning the Moon. Fourth, ascertain the state of the lunar interior. Finally, fifth, ascertain how the lunar surface features originated and evolved.

The ALSEP consisted of a central station, a radioisotope thermoelectric power source, and various experiment packages. The deployment of a typical installation is shown in figure 3.31. Experiments included active and passive seismic instrumentation, magnetometers, solar wind and mass spectrometers, ion detectors, a heat flow unit, charged particle detector, gravimeter, dust detector, and an instrument to detect lunar ejecta and meteorites.

The seismic stations provided important information about moonquakes and the internal structure of the Moon. Three primary types of moonquakes were identified: deep moonquakes that occur with remarkable regularity at the same locations and are most probably a tidal effect; shallow, irregular, and infrequent moonquakes occurring near the surface; and moonquakes caused by the impact of meteoroids. The internal structure of the Moon inferred from the seismic data indicate that any molten core must be smaller than a few hundred kilometers in radius; melting and differentiation of the outer crust probably took place, and this crust is two-layered and is mainly of anorthositic and gabbroic composition. Beneath the crust a mantle extends some hundreds of kilometers and shows a boundary region at which more iron may be present in the lunar material.

The magnetometers placed on the lunar surface provided information about the electrical conductivity, temperature, and structure of the crust and the interior, the

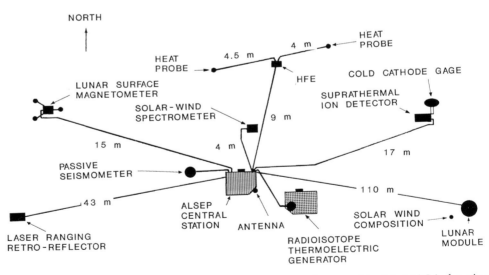

FIGURE 3.31 Deployment of a typical ALSEP package is illustrated in this NASA drawing of the Apollo 15 instruments and central station. Distance of each instrument from the central station is shown.

size of a lunar core and the abundance of iron in lunar material, the remanent magnetic fields, and the effects of the solar wind.

Free iron appears to exist in the lunar bulk in sufficient amounts to allow the Moon to be weakly ferromagnetic and not paramagnetic. The iron may be within a core or even as a shallow iron-rich layer. The magnetic permeability determined from this experiment suggests that while any core would have to have a radius of less than a few hundred kilometers, a core is not necessary to account for the data.

The crustal remanent magnetism might have been caused by thermoelectric effects early in the evolution of the Moon. If, for example, the infall of large bodies created mare basins, as is now generally accepted, and these basins filled with lava from sources within the crust, basins might be joined for a period beneath the surface by stretches of magma. These would effectively link together magma in the basins which could cool at different rates leading to different temperatures. The difference in temperature at each end of the connecting subsurface magma could lead to the generation of magnetic fields which might have been large enough to produce the remanent magnetism in the cooling lava of the basins and thus account for the magnetism measured in returned lunar samples. However, this origin for the fields is not generally accepted.

The ion detection experiment provided new information about the interaction of Earth's magnetotail with the Moon. An unexpected plasma regime consisting of low-energy plasma streaming away from Earth was discovered. It consists of oxygen and nitrogen ions from Earth's ionosphere.

The charged particle lunar environment experiment also provided important data about the distant regions of the Earth's magnetotail and magnetosheath, including the effects of geomagnetic storms. The instrument provided much new information about the interactions among photons and the lunar surface, solar and terrestrial plasma populations, and the environment on the Moon.

Gauges set on the lunar surface measured the minuscule amounts of gases present there. More gas was detected during the day than at night, but this was thought to be because of the astronauts' activity on the surface during daytime. The nighttime gas concentration was consistent with the neon expected to be deposited on the surface from the solar wind. The composition of this extremely rarefied lunar atmosphere was determined from a mass spectrometer experiment package. The most abundant gases are argon-40 and helium-4; the argon comes from radioactive decay of potassium-40 in the lunar interior. The helium source is the solar wind.

The solar wind spectrometer measured the properties of the solar wind impacting the lunar surface. The lunar ejecta and meteorites experiment recorded the transport of fine material of the lunar surface rather than incoming micrometeoroids.

The experiment to determine heat flow from the interior of the Moon also yielded important results. The experiment took place at two sites, Hadley Rille on the border of Mare Imbrium, and Taurus–Littrow valley. Most of the heat is believed to come from decay of radioactive isotopes and, if so, the heat flow measured would indicate that the uranium content of the Moon is proportionally about the same as that estimated for the mantle of Earth. Regional variations in heat flow, as measured at the two sites, probably arise from variations in the amounts of isotopes of thorium, potassium, and uranium.

Telemetry data from the ALSEP stations were received at Earth and recorded for 24 hours each day, amassing a tremendous amount of important information that scientists were able to evaluate long after the Apollo program ended.

• A NEW BEGINNING TO A GOLDEN AGE

The essence of Apollo was emphasized in comments by astronaut Eugene A. Cernan before the flight of the final Apollo, 17, when he discussed the design of the patch chosen for the mission.

"Our desire is that Apollo 17 symbolizes not the end of an era, but rather the culmination of the beginning of mankind's greatest achievements in his history—achievements which only have as their bounds the infinity of space and time—symbolization that man's seemingly impossible dreams can become limitless realities."

He further wrote: "We would like to recognize the historical foundation upon which the thoughts of the future are based—and, so as never to forget, we also hope to pay tribute to the Apollo Program and our nation, its people and its heritage, which have made these accomplishments all possible. The symbolism which captures these ideas sounds sophisticated and complicated to create. We hope it is not, because it is our desire to capture our theme with simplicity."

Robert T. McCall designed the Apollo 17 patch (figure 3.32). Its dominant design

FIGURE 3.32 Apollo emblems surround the emblem for the final mission, which the crew of Apollo 17 chose to represent not the end of an era but the beginning of a golden age of extraterrestrial exploration and development. In the emblem Apollo gazes toward Saturn and the Galaxy to symbolize that human aspirations should include the other planets and ultimately other star systems. (*NASA*)

element was Apollo, the Greek god of the Sun. Suspended in space behind the head of Apollo was a contemporary design of an American eagle the red bars of whose wings represented the bars of the American flag, and three white stars symbolized the three astronauts who formed the crew.

The background of the patch was deep blue space. Within it was the Moon, symbolically partially covered by the wing of the Eagle. Also the background included the planet Saturn and a spiral galaxy. The thrust of the eagle and the gaze of Apollo were toward other planets and the distant nebula. The implied message was that human goals in space someday will include the planets and perhaps the stars. The gold in the patch symbolized a forthcoming golden age of space flight continuing beyond the landing of the last Apollo.

The unmanned and manned missions to the Moon are summarized in table 3.3. Apollo astronauts—those greatest of explorers who were the first humans to travel

TABLE 3.3 *Lunar Explorations*

Launch Date	Name	Country	Results
1958			
August 17	Lunar Probe	U.S.	First U.S. attempt to fly by Moon; failed.
October 11	Pioneer I	U.S.	Failed to reach Moon.
November 8	Pioneer II	U.S.	Failed to reach Moon.
December 6	Pioneer III	U.S.	Failed to reach Moon.
1959			
January 2	Lunik 1	USSR	Lunar flyby that passed 3000 miles above the surface on January 4, 1959; no magnetic field detected.
March 3	Pioneer IV	U.S.	Passed 37,300 miles from Moon.
September 12	Lunik 2	USSR	Impacted Moon September 14, on Mare Imbrium; at 0° longitude and 30° N latitude; no lunar magnetic field was detected.
October 4	Lunik 3	USSR	Orbited Moon; first photos of far side.
November 26	Atlas–Able 1	U.S.	Failed at launch.
1960			
September 25	Atlas–Able 2	U.S.	Failed at launch
December 15	Atlas–Able 5B	U.S.	Failed at launch
1961			
August 23	Ranger 1	U.S.	Traveled only 312 miles from Earth.
November 18	Ranger 2	U.S.	Reached 146 miles from Earth
1962			
January 26	Ranger 3	U.S.	A lunar flyby which passed Moon at 22,862 miles and entered solar orbit.
April 23	Ranger 4	U.S.	Crashed into Moon.
October 18	Ranger 5	U.S.	Missed Moon by 450 miles, a flyby that entered solar orbit.
1963			
April 2	Luna 4	USSR	Missed Moon by 5300 miles.
1964			
January 30	Ranger 6	U.S.	First U.S. lunar impact on the Moon's surface, but no pictures were returned.
July 28	Ranger 7	U.S.	Successful impact on Mare Cognitum, at 20.6° W, 10.6° S; many pictures returned.
1965			
February 17	Ranger 8	U.S.	Impact on Mare Tranquillitatis; at 24.8° E, 2.7° N; good photos returned.
March 21	Ranger 9	U.S.	Impact in Alphonsus; at 2.4° W, 13.1° S; many good images returned.
May 9	Luna 5	USSR	Failed to soft land.
July 8	Luna 6	USSR	Missed Moon by 100,000 miles.
July 18	Zond 3	USSR	Lunar flyby returned photos of far side.

TABLE 3.3 *Lunar Explorations (Continued)*

Launch Date	Name	Country	Results
October 4	Luna 7	USSR	Lunar lander, impacted.
December 3	Luna 8	USSR	Lunar lander, impacted.
1966			
January 31	Luna 9	USSR	Landed February 3, in Oceanus Procellarum; returned photos.
March 31	Luna 10	USSR	Lunar orbiter
May 30	Surveyor 1	U.S.	Landed June 2, Oceanus Procellarum; at 43.4° W, 2.4° S, 35 miles N of Flamsteed.
August 10	Lunar Orbiter 1	U.S.	Orbited successfully; many photos returned.
August 24	Luna 11	USSR	Lunar orbiter.
September 20	Surveyor 2	U.S.	Failed
October 22	Luna 12	USSR	Lunar orbiter, photos.
November 6	Lunar Orbiter 2	U.S.	Orbited successfully; many photos.
December 21	Luna 13	USSR	Lunar lander, photos.
1967			
February 4	Lunar Orbiter 3	U.S.	Orbited, many photos.
April 17	Surveyor 3	U.S.	Landed April 20, in Oceanus Procellarum; at 23.3° W, 2.9° S.
May 4	Lunar Orbiter 4	U.S.	Orbited, many photos
July 14	Surveyor 4	U.S.	Failed
July 19	Explorer 35	U.S.	Orbiter
August 1	Lunar Orbiter 5	U.S.	Orbited, many photos.
September 8	Surveyor 5	U.S.	Landed September 11 in Mare Tranquillitatis; at 23.2° E., 1.5° N.
November 7	Surveyor 6	U.S.	Landed November 10 in Sinus Medii; at 1.5° W., 0.5° N.
1968			
January 7	Surveyor 7	U.S.	Landed January 10 in ejecta blanket of Tycho; at 11.4° W, 40.9° S.
March 2	Zond 4	USSR	No details available.
April 7	Luna 14	USSR	Orbited Moon.
September 15	Zond 5	USSR	Circumlunar flight and recovery.
November 10	Zond 6	USSR	Circumlunar flight and recovery.
December 21	Apollo 8	U.S.	First manned flight around Moon; no landing.
1969			
February 21	N1-1	USSR	Booster failed. *Note:* N1-1 through N1-4 were boosters for manned Soviet flight to Moon.
May 18	Apollo 10	U.S.	First lunar orbit rendezvous and close approach to lunar surface; no landing. *Note:* Apollo 7 (October 1968) and Apollo 9 (March 1969) were Earth orbit test flights.
July 3	N1-2	USSR	Failed on launch pad.
July 13	Luna 15	USSR	Lunar orbiter.
July 16	Apollo 11	U.S.	First human landing on Moon, July 20 in Mare Tranquillitatis; at 23.3°E, 0.6° N, in general area of Surveyor 5 landing and Ranger 8 impact.
August 8	Zond 7	USSR	Circumlunar flight and recovery.
November 14	Apollo 12	U.S.	Second manned landing November 19 in Oceanus Procellarum close to landing site of Surveyor 3; at 23.4° W, 3.2° S.
1970			
April 11	Apollo 13	U.S.	Aborted mission, flew around Moon; no landing.
September 12	Luna 16	USSR	Landed September 20 in Mare Fecunditatis; drilled into regolith and returned samples.
September 20	Zond 8	USSR	Photos of far side.
November 10	Luna 17	USSR	Landed November 17; Lunokhod 1 roamed on lunar surface for 11 months; returned samples to Earth.
1971			
January 31	Apollo 14	U.S.	Third manned landing on February 5, at Frau Mauro; at 17.5° W, 3.6° S in lunar highlands.
July 21	N1-3	USSR	Launch failure

TABLE 3.3 *Lunar Explorations (Continued)*

Launch Date	Name	Country	Results
1971			
July 26	Apollo 15	U.S.	Fourth manned landing on July 30 at Palus Putredinis near Hadley Rille; at 3.6° E, 26.4° N.
September 2	Luna 18	USSR	Lunar orbiter.
September 28	Luna 19	USSR	Lunar orbiter, photos.
1972			
February 14	Luna 20	USSR	Returned samples from lunar highlands; at 55° E, 5° N.
April 16	Apollo 16	U.S.	Fifth human landing, April 20 on Cayley Plains in lunar highlands of Descartes region; at 15.5° E, 9.0° S.
November 23	N1-4	USSR	Failed in late boost phase
December 7	Apollo 17	U.S.	Final Apollo, landed December 11 in Taurus-Littrow Valley floor; at 30.8° E, 20.1° N.
1973			
January 8	Luna 21	USSR	Lunokhod, landed January 16 in crater Lemognier, Mare Serenitatis; remote control surface exploration.
June 10	Explorer 49	U.S.	Radio astronomy beyond far side of Moon.
1974			
May 29	Luna 22	USSR	Orbited and returned photos.
October 23	Luna 23	USSR	Landed, but could not return samples; drill was damaged
1976			
August 9	Luna 24	USSR	Landed, returned samples from Mare Crisium.
1990			
January 25	Muses-A	Japan	Small lunar orbiter. Successfully entered lunar orbit March 19.

from Earth to another world (table 3.4)—spent a total of 30 full days in orbit around the Moon and a total of 12 1/2 days on the surface of the Moon. They had completed the preliminary exploration of Earth's satellite and laid the important groundwork of technology and science for humans to establish permanent outposts and bases on our neighbor world. The presence of humans beyond the confines of the planet on which they evolved had been demonstrated as possible. Whether humankind would accept the responsibility of adulthood as an interplanetary species or continue in its childlike territorial aggressions to control the ever-changing surfaces of its own planet remains to be seen. The portal to the future had been opened. But like Alice in a nonsensical wonderland of terrestrial politics and ideologies, many humans believed they could not fit or did not have the resources to go through that portal.

TABLE 3.4 *The Great Explorers*

FIRST HUMANS TO CIRCUMNAVIGATE ANOTHER WORLD

William A. Anders, Frank Borman, Eugene A. Cernan, Michael Collins, Ronald E. Evans, Richard F. Gordon, Fred W. Haise, Jr., James A. Lovell, Thomas K, Mattingly, Stuart A. Roosa, Thomas P. Stafford, John L. Swigert, Jr., Alfred M. Worden, John W. Young

FIRST EXPLORERS OF THE SURFACE OF ANOTHER WORLD

Edward E. Aldrin, Neil A. Armstrong, Alan L. Bean, Eugene A. Cernan, Charles Conrad, Jr., Charles M. Duke, Jr., James B. Irwin, Edgar D. Mitchell, Harrison H. Schmitt, David R. Scott, Alan B. Shepard, John W. Young

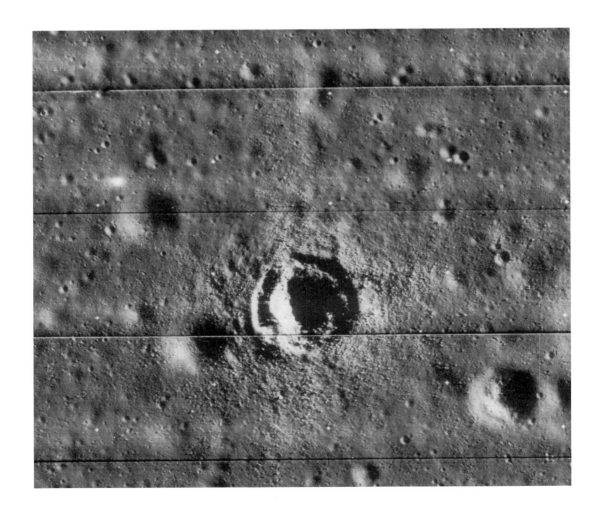

4 APOLLO'S MOON

Before the space program many questions about the Moon were waiting to be resolved. While this is no longer true, of course, the explorations raised many additional, and more complex, questions. The program results confirm not only that the dark maria were excavated by impacts and later filled by extensive lava flows, but also that most craters originated from impacts and are not volcanic. The surface layers of the Moon are regoliths of debris. Earlier speculative theories of "quicksand" of dust, deep layers of treacherous vacuum-fluffed unconsolidated rubble, and other exotic ideas proved erroneous. The surface debris is cohesive, analogous to damp sand, but while intricately molded and friable on the top surface, it is firm a few centimeters below that surface, much like a compacted fill.

Included in 849 pounds (385 kg) of lunar material collected by the Apollo astronauts were more than 2000 lunar rocks, nearly 100 soil samples, and several cores drilled from the lunar surface. A few grams of the material payloads returned to Earth by the Soviet unmanned sample return missions Luna 16 and Luna 20 were also available for study by U.S. scientists.

Analysis of the lunar rocks provided records of events in the Solar System far earlier than the history contained in Earth's rocks. The time when the Moon formed was established more closely; the absolute ages of subsequent events were measured, and the measurements confirmed the stratigraphic analysis of relative ages that had been made before the space missions. Two significant events are recognized; the fractionation of the crust and the mantle about 4.4 to 4.5 billion years ago, and the extrusion of mantle material onto the mare surfaces about 3.2 to 3.8 billion years ago. Absolute ages proved to be incredibly great, with little significant changes on the Moon since formation of the oldest terrestrial rocks we have been able to sample. Some evidence was obtained that the Moon formed by accretion, not necessarily from the solar nebula but possibly from materials shattered into space as a result of a Mars-sized body colliding with an accreting Earth. However, the Moon might have formed from a huge blob of material blasted into space as the result of an off-center collision of a planet-sized body with Earth. A mixture of material blasted from Earth and the impacting body might have entered an orbit around Earth where it coalesced to form the Moon.

Material produced by the subsequent cratering and basin creating-impact process was identified as resulting from different degrees of heating and cooling, related to the size of each impact and the location of the sampled material to the impact. The lunar material had been heated to various levels; some not high enough to melt, some high enough to produce glasses, and some to levels sufficiently high to melt the entire mass. Cooling rates also varied. Rapid cooling produced fine glassy fragments; slow cooling produced coarse-grained materials.

Mare basalts are relatively young and iron rich, in fact, richer in iron than terrestrial basalts, although the Moon generally is deficient in iron compared with the Earth. The highlands, by contrast, are relatively old and aluminum rich. Each had parent materials that were different and did not represent the original unmodified material of the Moon when it formed.

Gamma ray spectrometers, carried around the Moon in the Command/Service Modules of Apollo 15 and Apollo 16, recorded emissions from the Moon's surface and provided information on the amounts of potassium, uranium, thorium, iron, titanium, and magnesium in the surface swatches covered. Radioactive "hot spots" were discovered in the western parts of Mare Imbrium. "Cold spots" were located on the far side of the Moon.

The analysis of the samples and data from the various instruments confirmed that the Moon has an aluminum-rich refractory crust from which the lunar highlands evolved. Below this crust is an iron-magnesium-rich mantle, the most likely source for the dark basalts that flooded the maria and the floors of the larger craters. Below the mantle there might be a small core enriched with iron sulfide or iron, or both.

The Moon seems essentially different from other planetary bodies; it does not have the same chemistry as that of the Earth or of chondritic meteorites. The Moon seems to be short of potassium, enriched in rare-earth elements, and deficient in volatile precious metals such as gold, silver, and platinum. Isotopic composition of oxygen in lunar samples suggests that the Moon and Earth formed in the same general region of the Solar System.

• LUNAR ROCKS

From the Apollo experience we now know that the Moon has numerous rock types on both the highlands and the maria. A detailed knowledge of the petrology of lunar rocks was obtained, and their chemical compositions show that the lunar crust is depleted in volatiles and not as rich in potassium and sodium as is Earth's crust. It consists mainly of magnesium and iron silicates. The crust of the Moon is different chemically from the bulk of the continental crust of Earth, and it is also thicker than that of Earth.

While some mare basalt rocks have a high titanium content, they are in the minority. Some data have indicated that the Moon is enriched as a whole compared with Earth but nowhere nearly as much as the high-titanium mare basalts would suggest. The Moon has more refractory elements than does the Earth, and mare basalts are short of alkalis. The abundance of iron-rich and titanium-rich minerals, which are dark in color, accounts in part for the dark appearance of the maria.

The Surveyor spacecraft that landed close to the fresh crater Tycho, whose bright rays were spread widely and relatively recently across the face of the Moon, first indicated that the nature of the highland rocks was likely to be ancient lunar crust. Subsequent Apollo samples confirmed this. The rocks have high proportions of calcium-rich minerals and oxides of aluminum. They do not have as much iron as the basalts.

Ferroan anorthosites, the result of fractionation, are abundant plutonic lunar rocks prevalent in the highlands. They are, however, not primordial material although they are very ancient. Instead, they have turned out to be products from a magma ocean. More than 85 percent of their content is plagioclase feldspar. Anorthosites are formed as igneous liquid rocks cool; heavier crystals sink to the bottom of the molten mass while lighter crystals rise toward the surface. If lunar anorthosite is 30 miles (50 km) or so thick, as seems likely from the results of several experiments, these rocks must have been formed from an ocean of liquid rock encompassing the Moon early in lunar history.

Norites, troctolites, and gabbro-norites form the basis of what are called the "magnesium-suite" because they have high ratios of magnesium to iron and more magnesium than the ferroan anorthosites. These rocks are formed by intrusion into the anorthositic crust.

Alkali anorthosites are related to the magnesium-suite rocks but are richer in potassium and sodium than are the ferroan anorthosites.

Granites are rare on the Moon but are important rocks with quartz, potassium feldspar (orthoclase), and relatively sodium-rich plagioclase.

KREEP basalts are true extrusive rocks found in highland and mare areas. They are enriched in potassium (K), rare-earth elements (REE), and phosphorous (P). This rock type, too, is not primordial material. KREEP basalts contain much more plagioclase than do mare basalts and are the most terrestrial-like of all lunar basalts. These rocks were found mostly at the landing sites of Apollo 15 and Apollo 17, but samples were gathered at other landing sites also.

All lunar rocks are extremely old by terrestrial standards. Even the mare basalts are between three and four billion years old, and they are the youngest lunar rocks in the returned samples. The events that produced these lava flows lasted for almost one billion years, which is a longer time span than the period on Earth from the earliest abundant fossils at the beginning of the Cambrian era to the present day. During this period the surface of the Moon was violently active as lava flowed over vast areas, and large and small bodies continued to smash into the surface from space. Such time scales are enormous compared with human standards; terrestrial continents and oceans can be born, grow, and disappear within only 200 million years, and whole species become extinct within 100 million years. The dinosaurs in their great variety of forms, for example, became extinct only 63 million years ago.

Big lava flows are not unique to the Moon. There is evidence of lava flows on Earth as extensive as some of those on the Moon. Asish R. Basu and Paul R. Renne, writing in *Science* in 1991, reported that a lava eruption 250 million years ago covered at least 130,000 square miles (336,500 sq. km) of what is now Siberia. The surface of Mars also displays evidence of vast flows of lava, as does the surface of Venus as revealed by the Magellan spacecraft.

There are probably some even younger rocks on the Moon than those sampled by the Apollo teams. How much younger is unknown. Young lavas may be present, for example, in the Marius Hills where there are many volcanic features. Also, young flows appear to overlie more ancient flows of the Mare Imbrium (figure 4.1), and in Oceanus Procellarum there is a zone of basalts that may be relatively young also.

The age of the highland material has been determined from the rock samples. Additionally, the use of ratios of trace elements in highland material has helped to

FIGURE 4.1 Following the molding of the Moon's surface by bodies crashing into it from space, a period of volcanic activity produced extensive flows of very fluid lava. This picture shows ejecta from the crater Euler (top right) overlying the basalt flows of Mare Imbrium. Several flow fronts are clearly visible at the bottom of the picture. (*NASA*)

identify the components contributed by the infalling bodies that explosively excavated many highland craters. Several classes of materials from those ancient meteorites have been established. The fragments show higher proportions of rare metals than in the bulk of the Moon. The surprise is that these bodies did not have the composition of meteorites falling on Earth today. It is probable that they originated from a planet that had not separated into a mantle and a core. Possibly they originated from the collision of a large asteroidal body with the Earth.

Some analyses suggest that the lunar highlands were formed about 3.9 to 4 billion years ago, but this may be too young because of the averaging effects that subsequent impacts produce on the age-dating results. The ages of lunar rocks derived from the various methods of dating (e.g., rubidium-strontium, potassium-argon, and lead-uranium-thorium ratios) indicate that the lunar mass melted, fractionated, and formed crust very quickly after the bulk of the Moon formed some 4.6 billion years ago. The process appears to have been comparatively rapid with the crust forming within a few hundred million years of the Moon becoming a world. The results of one age-dating method suggest that major melting and separation of lunar rocks may have taken place somewhat later, 4.4 billion years ago, but this may be the time when the crust of the Moon solidified.

As mentioned earlier, lunar highlands (figure 4.2) are the oldest lunar rocks, an ancient crust pounded by bodies from space that blasted craters on top of earlier craters. The rocks are mainly anorthositic gabbros, an igneous rock consisting of pyroxene mixed with feldspar. Most samples from the highlands are 3.9 billion year old breccias—rocks that have been repeatedly broken by impact (figure 4.3). Breccias consist of mixed fragments of preexisting rocks that were "weathered" by the hail of meteorites from space. Plutonic rocks within the breccias are mainly anorthosites in which the proportion of iron oxide exceeds magnesium oxide. They include norites, dunites, and spinal troctolites. These rocks probably crystallized several kilometers deep within the crust of the Moon 4.3 to 4.6 billion years ago, and were later crushed and heated by the crater-producing impacts.

The breccia samples consist of coarse fragments of minerals and glasses set in a matrix of fine-grained, fragmental, or crystalline material. They are complex and contain more than one type of matrix and several types of fragments. The fine matrix reveals the process by which the rocks were broken. The fragments provide information about the original rocks.

These lunar breccias are of several types. Glassy matrices contain mineral, glass, and rock fragments within a matrix of the same material bound by grain-to-grain sintering. Crystalline matrix breccias consist of mineral and rock fragment clasts set in an interlocking matrix of crystalline minerals. They occur only in the lunar highlands. Additionally there are breccias similar to the glassy matrices, but lacking glass, and friable breccias consisting of crushed rock fragments; and there are rare breccias consisting of a tightly packed crystalline matrix and sparse mineral and rock clasts thought to have originated from deep in the highland crust. Criteria have been developed to distinguish igneous rocks from metamorphosed breccias and more than 100 samples of so-called "pristine" highland rocks have been identified.

Some KREEP basalts are believed to have been thrown about the surface of the Moon by meteorites excavating the highlands because, although KREEP is basaltic,

FIGURE 4.2 The oldest lunar rock are in the highlands. This oblique view of the far side shows typical rugged highland terrain of overlapping craters. The large crater with central peaks is 50 miles (80 km) in diameter. (*NASA*)

it is different from the mare basalts. Some KREEP basalt is thought to have originated from volcanic activity; other samples probably originated from local melts produced by the energy of impacting meteorites.

Crustal material spread over wide areas by impacts produced layers of debris that form analogs of terrestrial sedimentary rocks. Most complete rocks that are thrown onto the surface of the Moon are hit and shattered by a meteoroid within 100 million years. Metamorphosed rocks have been cycled through the brecciating process several times.

Generally it was thought from the Apollo results that the Moon became surrounded with a deep ocean of liquid rock soon after its formation. This ocean might have been as much as 300 miles (500 km) deep. The magma ocean hypothesis was based on the presence of anorthosite, which contains about 85 percent plagioclase feldspar. The question of whether or not this magma ocean did exist might be

FIGURE 4.3 This lunar rock sample is a polymict breccia. Its matrix consists of crushed minerals; this sample is generally free of glass. Breccias are equivalents of terrestrial sedimentary rocks; they consist of earlier rocks that have been weathered by impacting meteorites and recompacted, sometimes metamorphosed into new rocks. (*NASA*)

settled if we knew how much of the Moon's crust consists of anorthosite. The question is important to interpreting the early history of our own planet: if the Moon possessed a magma ocean soon after its formation it would be reasonable to assume that the Earth possessed one also.

A SURFACE POUNDED INTO SOIL: THE LUNAR REGOLITH

Draped over the hard rock of the entire lunar surface is a regolith, a layer of unconsolidated debris generated over billions of years by the pounding hail of meteoroids. This lunar soil is different from terrestrial soils because it was not subjected to interaction with water, an atmosphere, or living systems. Terrestrial soil originates from the actions of planetary processes of weathering and of the biosphere. Lunar soil is generated by cosmic processes, the infall of particles and radiation from space. The surface of lunar soil is also saturated with gases from the solar wind, and as a result it contains a record of the flow of charged particles from the Sun.

Thickness of the regolith varies. At the edge of the Hadley Rille on Palus Putredinis, for example, astronauts discovered that the regolith layer is so thin that bedrock is exposed. The layer is also thin on the flanks of young craters. At the Apollo 16 site on the Cayley Plains of Descartes region, a crater that has been

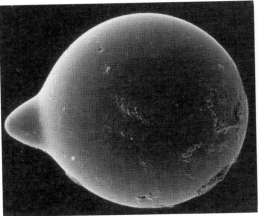

FIGURE 4.4   The Moon differs from Earth in that there are many glasses in lunar rock. On Earth glass loses its glassy properties rapidly because so much water is present. On the Moon there is virtually no water, so glasses are ubiquitous. They were probably formed by heat generated from impacts. (a) Micrograph of a spherule of lunar glass. (b) An exciting discovery at Taurus-Littrow by geologist Schmitt was of orange colored soil (the light area just below the tripod) consisting almost entirely of glass beads that are thought to have originated from lava fountains at the rim of the Serenitatis Basin. (*NASA photos*)

named the North Bay Crater is about 50 million years young. On the rim of this crater the regolith is only a few centimeters thick. Elsewhere, as is evidenced where craters have penetrated underlying bedrock, the regolith is several yards (meters) thick. It is 13 to 17 feet (4 to 5 meters) thick on the mare surfaces and up to twice that thickness on the ancient highlands. The additional regolith on the highlands compared with the maria probably resulted from the impacting rate of meteoroids being higher in the period before lava flooded the mare basins than it was afterward.

Cores from several Apollo landing sites reveal that the regolith has stacked layers of ejecta from craters, each layer being well mixed on a small scale. The layering from core to core differs, showing that it is purely local. The upper few millimeters of the regolith are turned over hundreds of times in a million years, whereas a meter or so below the surface the soil remains undisturbed for hundreds of millions of years.

Soil chemistry at any locality generally represents the underlying bedrock, but other materials are introduced from great distances by ejecta spread by major impacts. The material of the incoming bodies also adds to the regolith. Although some materials are exchanged between highland and mare surfaces, there is no large scale transport of material from one to the other.

Impacts cause glass, pyroxene, and olivine fragments to become concentrated in the finer size particles, whereas plagioclase minerals are in coarse fragments. Some elements are thereby enriched in the fine grained material, notably strontium, barium, rubidium, and potassium.

The lunar soil contains many glass spherules (figure 4.4a), including emerald green glasses at the Apollo 15 landing site on Palus Putredinis, and orange glasses (figure 4.4b) discovered by the sole geologist/astronaut, Harrison Schmitt, at the Apollo 17 landing site in the floor of Taurus-Littrow valley. These glass particles are in many forms; some are dumbbell shaped, others teardrops, and yet others are jagged fragments. Some glasses are believed to have been formed by volcanics and others by the heat generated by impacting bodies. The orange glass spheres gathered at Taurus-Littrow are 3.8 billion years old. They were excavated from within the mare basalt when the impact that produced Shorty Crater penetrated through the material of an ancient landslide in the valley where Apollo 17 landed. Their discovery is a good example of the importance of geologists in field work on the Moon.

Scientists speculate that these glass beads, which contain high proportions of volatiles, originated from ancient lava fountains at the edges of the mare basins. Showers of liquid lava at low viscosity shot high above the lunar surface from vents on the edges of the cooling lava fields, and the escaping volatiles produced innumerable droplets that cooled into glass spheres.

Those glasses on the Moon generated by the heat of meteor impacts were produced as the molten rock rapidly cooled after it splashed from each impact site, like droplets of mud from a mud hole hit by a thrown rock. The heating effect of meteorite bombardment also caused glasses to bind together soil particles previously broken down by the processes forming the regolith. Within the regolith there has always been both building and disruptive processes at work.

• BULK GEOCHEMISTRY

The previous chapter discussed how Apollo investigated the bulk geochemistry of the Moon by observations from orbit and by experiments on the surface. An X-ray fluorescence experiment from orbit measured X-rays generated in the topmost layer of the surface by X-rays from the Sun. The experiment established ratios for magnesium, aluminum, and silicon along narrow swathes of surface below orbiting Command Modules of some Apollo missions. The highlands are richer in aluminum than the maria. The ratio of aluminum to silicon in the highlands matches that of the igneous rock, anorthositic gabbro.

While the flooding lavas probably developed from partial melting of the lunar interior there is only some suggestive evidence of large-scale fractionation of lavas near the surface; no equivalent of highly fractionated rocks of terrestrial lavas, such as those of Hawaii. Fractionation occurs when liquid rocks cool slowly enough for mineral crystals to separate in groups—lighter minerals floating and heavier minerals sinking.

With the fluidity of heavy oil, lunar lavas flowed freely. Texture of lava samples suggests that the lavas cooled rapidly and crystallized quickly without changing their chemical composition, as would have resulted from separations of early formed mineral phases. But some lavas in deep pools might have taken longer to cool and could have changed in chemical composition during the cooling process. If such pools can be located they may become an important source of minerals for development of the lunar base.

The lunar samples suggest, however, that the Moon has comparatively few mineral species—only 100 compared with 2000 terrestrial minerals. This is probably because primitive lunar material was unable to react with an atmosphere or with water.

The number of primary materials (minerals that crystalize from silicate rocks) is similar on both Earth and Moon. Principal lunar minerals are pyroxene (a calcium-iron-magnesium silicate), plagioclase feldspar (a calcium-aluminum silicate), olivine (an iron-magnesium silicate) and ilmenite (an iron-titanium-oxide). The last named may be the source of oxygen for the lunar base when it is established.

Lunar pyroxenes, even in single crystals, show greater variations than their terrestrial counterparts. Lunar plagioclase is richer in calcium than its terrestrial varieties. Some samples are almost pure feldspar; on Earth the pinkish mineral feldspar is found in granite. It is an end-product of a complex process of evolution from igneous rocks. The mineral seems concentrated on the lunar highlands to produce anorthositic gabbros, coarse-grained igneous rocks similar in composition to basalts.

Olivines crystalize early from hot magma. In the lunar basalts most are rich in magnesium with some rich in iron. Lunar ilmenite is nearly pure iron-titanium-oxide.

Geochemically, lunar samples are characterized by many refractory chemical elements such as uranium, thorium, and zircon, and rare-earth elements. These elements cannot become incorporated significantly into the rock-forming minerals commonly found on Earth. On the Moon they produced unusual mineral phases leading to many accessory minerals that complicate the study of lunar mineralogy. Unlike terrestrial rocks, lunar rocks contain no traces of water.

FIGURE 4.5   This Apollo 15 sample is a highly vesicular fine-grained basalt collected by the crew about half a kilometer from the Hadley Rille. The vesicles are smooth walled and spherical. (*NASA*)

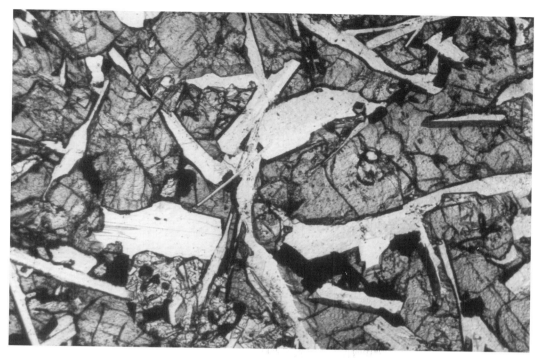

FIGURE 4.6　Micro section of lunar sample viewed under a polarizing microscope shows details from which much can be learned about the history of the lunar rocks and when the different parts crystallized. (*NASA*)

Samples from lunar maria contain vesicular basalts (figure 4.5), rocks in which entrapped gas bubbles produced bubble pits. Minerals often crystallized on the inside of the vesicles as the rock cooled. The chemically diverse mare basalts are strongly reduced relative to their terrestrial counterparts since they were formed under conditions where free oxygen was scarce. This led to lunar basalts containing metallic iron—iron sulfide—rather than ferric iron. Chemical differences between the lunar and terrestrial basalts indicate that the interior of the Moon has a different composition from the mantle of Earth.

Highland rocks are not as clearly defined geochemically as are the rocks of the maria. They have been so badly damaged and reworked by impacts that their original igneous nature is partially masked. On the average they have greater proportions of feldspar than do mare rocks. Today, almost all the highland rocks are breccias; mixtures of earlier rocks broken by meteorite impacts and later metamorphosed and refabricated by subsequent impacts.

When sliced extremely thin, some lunar rocks are transparent to light and can be examined in a microscope as thin sections (figure 4.6). Such sections yield many details about the rock: the minerals present, their arrangement, shape, and clues as to their sequence of formation. For example, a mass of grains or minerals surrounded by another mineral indicates that the surrounding mineral crystallized after the small grains within it.

• PHYSICAL PROPERTIES OF THE MOON AS A WORLD

The physical properties of the lunar surface and interior reveal information about the structure, composition, and state of the Moon. This information is important not only to provide new insights into the geology and geophysics of the Earth but also to determine the origin and evolution of other planets and of the Solar System in general.

The mass and radius of the Moon and its mean density of nearly 3.5 times that of water (3.34 g/cm^3) were known with a high degree of precision before the era of space exploration. However, the moments of inertia were not accurately known, and consequently the arrangement of materials of differing densities within the Moon was questionable.

If the Moon were a homogeneous body, as was at one time generally believed, its central pressure would be less than if it has a dense core. One of the major tasks of lunar geophysics was to determine if the Moon differentiated so that today it possesses a core.

Measuring the heat flow from the interior is important since it can reveal information about the internal temperature and the conductivity of the interior, and provide clues about the thermal history of the Moon, which is, of course, vital to understanding its evolution. Since it is believed that the lunar heat originates from uranium and thorium concentrated in the crust, the flow rate permits calculation of the amount of these elements. But limits are placed upon these amounts by the measured abundance of uranium and thorium obtained from the lunar rock samples. Unfortunately the two experiments made during the Apollo missions, as described in the previous chapter, did not yield conclusive evidence about average global heat flow to the lunar surface. Each instrument was emplaced at the edge of a mare basin and might have been recording unusual heat flow activities at such a location. A surprisingly high result of 0.74 heat flow units was recorded at the Apollo 15 Hadley Rille site. The average on Earth is only 1.5 heat flow units, about twice the lunar value. At the Apollo 16 Descartes site, the equipment failed and no data were obtained. At the Apollo 17 site in Taurus-Littrow Valley the lunar heat flow was similar to that measured at the Apollo 15 Hadley Rille site. These results raise doubts about the validity of the Apollo measurements being applied to the whole of the Moon.

That the Moon is asymmetric has long been known, but it was not until spacecraft orbited the Moon that its precise figure was determined. The most accurate measurements were made from Apollo Command/Service Modules in orbit about the Moon. These used first radar and then laser altimeters to measure accurately the distance between the spacecraft and the lunar surface and thus determine how the shape of the surface varies relative to the spacecraft's orbit about the center of mass of the Moon. As mentioned in chapter 2, the center of mass is offset from the center of the geometrical figure by 1.24 miles (2 km) in the direction of the Earth but slightly toward the east. If the Moon has a core, the gravity data can be satisfied by assuming that the core is displaced toward the Earth, while on the far side of the Moon the crust is considerably thicker than on the maria of the near side. This could account for the maria being filled on the near side and not on the far side, because the warm, source region of lunar basalts was closer to the surface on the Earth-facing hemisphere. When the bodies that impacted to form the basins of the maria hit the Earth-facing side they were able to disturb the crust down to the level of the basalt source, whereas similar bodies impacting on the far side of the Moon were not able to do so.

This offset and asymmetry has been used to argue that the Moon is a cold rigid body. However the stresses to hold the Moon in its asymmetric shape are fairly small compared with the strength of the thick lunar crustal rocks and the underlying mantle. The thickness of the crust may vary from 20 miles (30 km) on the near side to more than 100 miles (150 km) on the far side. Calculations show that a cold crust and lithosphere could maintain the distorted shape over a hot asthenosphere.

The laser altimeter showed that the highlands are about two miles (3 km) above the level of the maria. While the mare surfaces are relatively smooth and level, the highlands vary in elevation over a range of about two miles (3 km) from their mean level.

Radio tracking of lunar orbiting spacecraft provided a first detailed analysis of the lunar gravitational field. Gravitational anomalies associated with the large lunar basins such as Mare Imbrium and Mare Serenitatis were discovered. Other circular maria show anomalies: the filled maria have heavier material within the basin and less dense material surrounding it. Unfilled or partially flooded basins, such as Mare Orientale, have less dense material within them.

Several suggestions have been put forward for the origin of these positive gravitational anomalies called mascons that indicate heavier rocks within basins. Some explain the mass concentration was caused by a dense impacting body that contributed the added mass. Others explain the mascons as in some way due to the subsequent flooding of the basin with dense lava. It is unlikely that the incoming body would provide the mass for the mascons because the present gravitational anomaly is spread over a large flat area. More likely the mass arises from basaltic flows concentrating heavier rocks in the mare basins at the expense of the surrounding areas.

Such molten lavas coming from the upper mantle of the Moon would have permeated the zone of rubble produced by the impacting bodies and consolidated that rubble. It would, then, through a series of thin flows, gradually have filled the basin itself (figure 4.7). Although there are alternatives, this seems the most likely explanation of the mass concentrations. The negative anomalies surrounding such basins are explained by depletion of material around the basins as lava flowed into them. Unfilled basins, by contrast, show a surrounding ring of heavy mass and a shortage of mass over their central plains because material did not flow to fill the basins.

The presence of mascons provides important information about the Moon itself. The flooding of the maria took place over three billion years ago, and the mascons were presumably caused by this flooding rather than by the impact itself. Although mountain rings have sunk into equilibrium in what must have been a plastic crust—that is, a crust hot enough to deform at the time of the impact—the mascons have remained. This is interpreted as meaning that the Moon was sufficiently rigid to

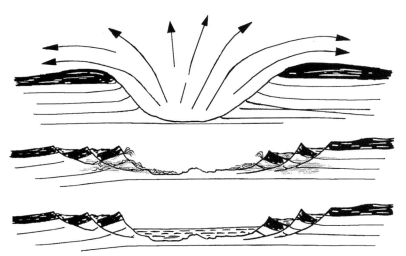

FIGURE 4.7   The ejecta from an impact spread round the basin and formed an insulating layer. Heat from radioactive elements melted rocks beneath this layer. The resulting lava drained through faults and crevices in the surrounding rubble to fill the basins densely with a series of flows. Along the basin rims, lava fountains gushed into space and formed glass droplets. The heavier rocks filling the basins gave rise to the mass concentrations (MASCONS) that produce the gravity anomalies over the basins.

support the mascons by the time the lava flowed, although it was not so rigid at the time the basins were excavated and the concentric ring mountains were formed.

• PROBING DEEP WITHIN THE MOON'S INTERIOR

Seismic experiments are another way to probe the Moon's interior. The most direct means by which geophysicists determine the internal structure of the Earth is through observations of seismic waves generated by earthquakes. An important question asked before the Apollo missions was whether the Moon is subjected to moonquakes. If there were seismic activity on the Moon, analysis of seismic waves would reveal the inner structure. Also anomalies in the lunar gravitational field could reveal the presence of material of different densities within the Moon. As detailed in the previous chapter, Apollo established a seismic network on the Moon, with stations placed by Apollo 12 and subsequent missions. The seismic network was used both passively to record natural seismic events and actively to record events triggered artificially or by the impact of meteorites.

Lunar seismograms differed markedly from their terrestrial counterparts. The moonquakes are of long duration and reverberate. They do not have well-defined surface waves, and their unique form may result from the extremely dry nature of lunar rocks. Moonquakes, of which several thousand occur each year, fall into two major categories. Many are deep events that take place at 400 to 600 miles (600 to 900 km) at about 50 identified centers. As recorded during the Apollo program, they occurred in a repetitive fashion governed by the monthly Earth tidal cycle and, over a longer period, by the solar tidal cycle. Shallow events had no particular pattern; they were much fewer than the deep events, and seemed to be related to stress being released in the crust of the Moon. In addition there were many microquakes close to the surface that were probably caused by rocks cracking or slumping at sunrise and sunset when surface temperatures change rapidly.

Seismic experiments provided information about the composition of the outer 100 miles (150 km) of the Moon. As mentioned in the previous chapter, some experiments directed toward probing the upper layers of the Moon used a thumper held by an astronaut; later ones used mortar-launched grenades. The impact on the lunar surface of spent Saturn IVB stages and the Lunar Module ascent stages generated strong vibrations that probed deeper into the Moon than those from the thumper and the impact of small meteorites.

The outermost 0.6–mile (1–km) thick layer of the Moon is composed of lightly compacted rubble rather than hard rock. This causes the seismic vibrations to be transmitted at very low velocities from their source to the seismometers. Below this unconsolidated layer the vibrations travel more quickly in what appears to be consolidated rubble or a rock that is characterized by many fractures. Still deeper, below 15 miles (25 km), the velocity of transmission increases rapidly and then remains almost constant to a depth of 47 miles (60 km) on the near side and 62 miles (100 km) on the far side. This velocity is consistent with anorthositic gabbros. At these depths another change occurs that seems due to the presence of a mantle consisting of rocks such as equivalents of terrestrial pyroxenites. This material appears to extend to a depth of 90 miles (150 km). There is also some evidence pointing to a high velocity layer at a depth of about 40 miles (60 km), and there is some evidence of a discontinuity at a depth of 300 miles (500 km). In the passive experiments the network detected many very small moonquakes, but the Moon is nowhere nearly so seismically active as the Earth. Also most moonquakes come from depths of hundreds of miles, while earthquakes originate at depths ranging from less than 12 miles (40 km) in California to 125 miles (200 km) when associated with ocean trenches.

Because the seismic wave velocity changes below the crust the rocks there must be denser than the crustal rocks; this effect is most probably from a lunar mantle. The mantle and the crust thus form a lithosphere that is probably about 600 miles (1000 km) thick. It is at the boundary between the lithosphere and an asthenosphere that the deep moonquakes originate. However, there is some evidence to show that this boundary is not regular.

While the lithosphere seems rigid, the asthenosphere is probably plastic. Lunar scientists have not yet determined whether there is a small iron-rich core within the asthenosphere and how hot it is. The induced magnetic field measurements neither confirm nor rule out the possibility of a small, iron-rich core of less than 300 miles (500 km) radius. What does seem clear, however, is that any active processes continuing within the Moon today must be confined to the asthenosphere by the rigid shell of the lithosphere. The thick outer shell of the Moon prevents any further movement of lava to the surface and jostling of plates of crust in tectonic activities such as those occurring on Earth today.

The form of moonquakes probably means also that the interior of the Moon is not in dynamic global convection as is the interior of the Earth. It seems that the lunar interior has an asthenosphere, or central region, on top of which the moonquakes occur (figure 4.8). There is a gradual transition to a surrounding lithosphere or mantle that is topped by 40 miles (60 km) of lunar crust on the near side and about twice this depth on the far side. The top 6 to 12 miles (10 to 20 km) of this crust is intimately mixed by the action of bodies impacting from space. The Moon's lithosphere is very thick and quite stable, whereas Earth's lithosphere is relatively thin and very active.

Because no shallow moonquakes were recorded, scientists believe that the heat flow from the interior of the Moon is quite regular in balancing the heat generated within the Moon. As a world, the Moon is thus neither expanding nor contracting, but is highly stable. One of the great surprises was that there is no detectable moonquake activity along the many straight rilles. They must have formed early in lunar history, and today they are no longer active fault zones.

However, because seismic shear waves do not seem to pass through the interior of the Moon below about 600 miles (1000 km) the asthenosphere may still be hot and some parts may be molten. There may, indeed, be a molten core with a radius of less than 225 miles (360 km). For molten material to exist today in the core of the Moon, radioactive sources must have remained at depth and not have floated to the crust. Alternatively, the Moon must be extremely viscous to reduce the convection of heat from the interior to the surface. One of the outstanding questions is whether the deep interior of the Moon is hot. Heat from an initially hot Moon could have been transported through conduction and convection to the surface and would have allowed the Moon to cool. However, conduction alone could only cool the outer 180 miles (300 km). The mare basalts also could have carried considerable quantities of heat from within the Moon to the surface as the lavas rose from depth, but it is unlikely that they originated much deeper than 300 miles (500 km). The floating of an ancient crust would have placed an insulating blanket on the hot interior and prevented further heat loss on a large scale. Below a thick lithosphere needed to support the mascons for more than three billion years, the asthenosphere could be molten today with its heat only very slowly escaping toward the surface.

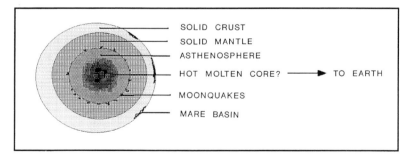

FIGURE 4.8   The lunar interior is believed to be mainly solid, with possibly a small, partially liquid core. Moonquakes occur at great depths at the boundary between the asthenosphere and the mantle. The crust of the Moon is believed to be thinner on the Earth-facing hemisphere than on the far side, thus accounting for the center of mass of the Moon being displaced toward the Earth.

• PAST LIFE OF THE DEAD MOON

Any theory for the thermal evolution of the Moon has to consider the scientific discoveries resulting from the Apollo program and the exploration of other planets of the Solar System and their satellites.

An interpretation of how the Moon evolved thermally is constrained by many factors that have been established as a result of measurements made by various instruments during the lunar science program. These factors include the unexpected high heat flow, the dates of lunar events such as general cratering, basin formation and filling, processing of highland crust by impact cratering, moments of inertia of the Moon, seismic activity, remanent and transient magnetic fields, distribution of lunar minerals, and the enrichment and depletion of various elements in the lunar rocks.

The Moon reached its present quiescent state after a period of intense activity billions of years ago. The story told by the rocks together with an analysis of lunar samples, coupled with many experiments on the surface and from orbiting spacecraft, contributed to a more realistic account of how the Moon might have evolved after it formed some 4.6 billion years ago.

The presence of some KREEP basalts originating 4.4 to 4.5 billion years ago requires an early period of lunar volcanism which, in turn, implies that the lunar interior had then to be hot enough to melt, at least partially, a large amount of plagioclase. Moreover, the source of the KREEP basalts had to be enriched with uranium at least five times over that found normally in chondritic materials. This older KREEP is believed to have originated in crystallization of a magma ocean.

The flooding of the mare basins required extensive volcanism which, in turn, required a partially molten interior to the Moon as recently as 3 billion years ago. There is some evidence that lava flows may have continued in Oceanus Procellarum to 2 billion years ago. When the lunar heat engine shut off, volcanics essentially ended. The outer several hundred miles of the Moon have been rigid enough to support the mascons for 3 billion years, possibly for as long as 4 billion years, which requires that the lithosphere of the Moon must have been relatively cool all this time. Today the interior of the Moon is substantially below the melting point of silicates, and yet the heat flux is high, implying that uranium is concentrated in the outer shell of the Moon.

Evidence obtained from lunar science provides what is believed to be a reasonable picture of what the outer 600 miles (1000 km) of the Moon might have been soon after its formation. It is important to note, nevertheless, that the Apollo missions explored only a very small sampling of the Moon. There are much more speculative theories about the innermost part of the Moon. One model suggests that the core region originally consisted of primitive unfractionated material in which some radioactive elements caused a partial melting but not enough for differentiation to take place and a metallic core to form. Another model assumes that the central regions melted and differentiated so that metals, including the siderophiles, collected in a core of predominantly iron and iron sulfide.

One thermal model of the Moon that allows for the seemingly conflicting requirements of differentiation of the outer 90 miles (150 km), partial melting of a core at 600 miles (1000 km), high global heat flow, survival of mascons for several billion years, and delay in the filling of the maria, relies upon the assumption discussed earlier that the Moon formed cold from many cold planetesimals. As the aggregation of planetesimals grew in size, the kinetic energy of the additional accreting bodies moving at higher velocities, under the influence of the increasing lunar gravity, provided a pulse of thermal energy sufficient to bring the surface to the melting point but not enough to penetrate deep into the interior. The sequence of evolution might have been as follows. The energy of accretion was great enough to cause much of the outer layers of the Moon to melt, thus allowing at least partial differentiation and separation of a crust from an ocean of hot rock on which the crust floated. This crust gave rise to the anorthositic gabbros dated as old as 4.6 billion years. How far into the Moon the melting progressed is not clear, but if the remanent magnetic field came from the Moon itself, the melting must have proceeded sufficiently to concentrate iron in the core and produce convection that allowed a planetary dynamo to occur before the surface rocks cooled and locked in the magnetic field.

These outer parts then cooled rapidly enough to record the impacts of the tail end of the accretion. Subsequently some large bodies that had accreted in the inner Solar System collided with the Moon (and, of course, with Earth) to produce the large basins. The impacts spread radioactive elements around the peripheries of the basins within a thermally insulating blanket of ejecta. Radioactive heating below this blanket was sufficient to melt rocks to provide the lavas that flowed into the basins from the surrounding deep crust. By contrast, the impact site itself was cold and rigid enough to accept the basalt flows and to develop a gravity anomaly. Only comparatively recently, after this episode of mare formation, has radioactive heating become great enough to melt partially the deep interior.

It seems that there was a stage of volcanism during which the crust melted above 60 miles (100 km) to produce a younger KREEP at about 4 billion years ago. This may have been in connection with the excavation of the impact basins, and energy for it may have come from the impacts of the large bodies that blasted the surface.

Extensive cratering also took place prior to and during this period—cratering that might have been the tail-end of accretion of Solar System bodies and in part associated with some Solar System cataclysm that produced the large bodies responsible for the impact basins.

Another period of volcanism took place from 4.0 to 3.2 billion years ago when the upper mantle melted and produced a series of lava flows (figure 4.9) with the lava rising from 90 to 150 miles (150 to 240 km) beneath the lunar surface, probably because of radioactive heating. Following this period, the Moon remained undisturbed except for a few large impacts that produced the bright ray craters and many smaller impacts that pockmarked the surface down to microscopic levels.

• A COMPASS IS CONFUSED ON THE MOON: LUNAR MAGNETISM

Apollo experiments measured two types of magnetic fields: transient fields that are developed by electric currents generated in the interior of the Moon, and permanent

fields that are due to fossil magnetism. Transient fields arise from the solar wind. Change in the magnetic fields carried by the solar wind induce electric currents within the Moon that, in turn, produce magnetic fields. These fields allow scientists to compute the electrical resistivity of the interior of the Moon and from this deduce the temperature of the deep interior. The transient fields also allow an estimate to be made of the amount of free iron within the Moon.

Simultaneous measurement of the field of the solar wind by an Explorer satellite orbiting the Moon and of the response of the Moon to the wind as measured by magnetometers placed by astronauts on the lunar surface, allowed the conductivity of the Moon to be calculated. An important result was that the electrical conductivity of the Moon rises rapidly to a depth of 155 miles (250 km) and then remains essentially constant to a depth of 480 miles (800 km).

The electrical conductivity of minerals that are believed to be present at these depths within the Moon is known. This allows the temperature of the interior to be calculated, since the conductivity of the minerals is governed by the temperature. Measurements for olivine, for example, can be used to convert conductivity into expected temperature and they show that permanently magnetized material could exist at shallow depths only, because for iron the Curie point (temperature at which magnetism can be frozen into the rocks) occurs at a depth of only 120 miles (200 km).

There is a rapid increase in internal temperature of about 49.1°F per mile (6°C per km) to 150 miles (250 km) and almost uniform temperature from that depth to the greatest depth for which the conductivity experiment provided data, namely to 500 miles (800 km).

Thus it seems that the temperature of the Moon is only about 2730°F (1500°C) at a depth of 600 miles (1000 km), and 1520°F (825°C) at a depth of 125 miles (200 km). The Moon is, therefore, relatively cool compared with Earth, and this is consistent with the low level of lunar seismic activity.

The relative magnetic permeability of the Moon can be derived from the magnetometer experiments, and the abundance of iron is calculated from this permeability. The abundance of free iron is about 2.5 percent by weight. This corresponds to a total iron abundance of 5 to 13.5 percent by weight for olivine and orthopyroxene respectively. Since the Moon is probably composed of a combination of these minerals, the total abundance of iron is probably somewhere between the above two extremes.

Magnetic field measurements made from Lunar Orbiter, Luna 10, and Explorer 35 indicated that the permanent magnetic dipole for the whole Moon was less than 1/100,000 that of the Earth—a vanishingly small field. It was because of these measurements that Apollo 11 carried no magnetometer. However, remanent magnetism was unexpectedly discovered in the samples. A breccia fragment returned from the Apollo 11 landing site possessed a strong and stable remanent magnetism, the discovery of which renewed interest in lunar magnetism, so that a magnetometer was carried to the lunar surface by the crew of Apollo 12. Subsequent missions also took magnetometers to the Moon. Strong surface fields were measured, for example, 103 gamma at the North Bay Crater at Descartes, but still weak compared with the Earth's field of 30,000 gamma.

Subsatellites (figure 4.10) launched from Apollo 15 and Apollo 16 Command Modules carried a magnetometer to measure the field from close lunar orbit. Local anomalies were recorded and the remanent magnetism was shown to be not merely a local effect concentrated fortuitously at all the Apollo landing sites, but also of global extent and varying in direction and magnitude.

Evidence of this remanent magnetism collected at the surface and from orbit indicated that, prior to 3.2 billion years ago when the lunar rocks cooled and locked in the field, the Moon had been subjected to a strong magnetic field of a few thousand gammas intensity over at least half a billion years. It is not known whether this field was generated by the Moon or was an interplanetary field in which the Moon was immersed at that time.

The experience with Apollo has added another mystery to the involved magnetic histories of planetary bodies. Once only the Earth and Jupiter were known to have

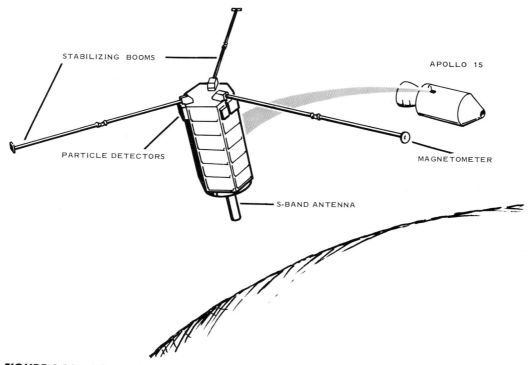

FIGURE 4.10 Three science experiments were included on each subsatellite launched into orbit around the Moon from Apollo Service Modules starting with Apollo 15. These experiments mapped the lunar magnetic field; also the gravity field (by S-band tracking) and the flux of energetic particles. (*TRW*)

significant fields and the simple dynamo theory seemed adequate to explain them. Now with a significant field detected on small, slowly rotating Mercury, while the slowly rotating Moon, with only at best a small core, also shows a remanent field, the whole theory of planetary magnetism is having to be restated. The exploration of the outer Solar System and the discovery of offset and tilted magnetic fields on Neptune and Uranus and a field on Saturn have also added more data points that have to be incorporated in magnetic theories for planets and their satellites.

• ALMOST A VACUUM; THE ONLY WIND ON THE MOON COMES FROM THE SUN

Astronomers had known for centuries that the Moon has virtually no atmosphere. The limb of the Moon appears sharp and clear with no trace of atmospheric softening. When the Moon passes in front of stars so that they are occulted, their light is cut off abruptly, again showing that the Moon is airless. The boundary between day and night, the terminator, shows objects casting long shadows with everything very clearly defined. There are no traces of atmospheric haze (figure 4.11).

FIGURE 4.11 Sharp shadows cast by lunar mountains and craters show that the Moon has no atmosphere of any consequence. There are no hazes to obscure the fine details of the surface revealed in photographs from orbit. (*NASA*)

Several observers as recently as the 1950s claimed that they had seen evidence of mists or hazes on the Moon. Some claimed to have detected the spectra of faint emissions within craters in what are referred to as lunar transient events. In 1959 the Soviet astronomer N. A. Kozyrev claimed to have obtained a spectrograph showing a gas emission in Alphonsus and, in 1969, a red glow in the deep young crater Aristarchus. This crater was also seen to brighten suddenly under Earthshine. More than 600 observations of lunar transient phenomena had been recorded since 1540. The author attended many of the series of Lunar and Planetary Exploration Colloquia held in the Los Angeles area during the late 1950s and early 1960s. He remembers Jack Green presenting a long and detailed list of many transient events. An involved and heated discussion among astronomers followed. A variety of speculations concerning these lunar events and the possibility of sporadic emissions of gas from the lunar interior suggested ways in which they might occur. Could they be the remains of an extremely diffuse atmosphere formed at or soon after the Moon's formation? This was most unlikely because normal atmospheric gases could easily escape into space; the escape velocity at the Moon's surface is only 1.46 miles per second (2.35 kps). Were they caused by long-term volcanic activity continuing to outgas volatiles, or possibly the volatization of surface materials by the heat generated from the impact of meteorites? Whatever the origin the consensus was that the total amount of atmospheric gases on the Moon could not amount to more than a few tons. This type of atmosphere could obviously be overwhelmed by the rocket exhaust gases released on the Moon during the Apollo program.

The experimental upper limit on the pressure of a lunar atmosphere at the surface was estimated as one ten millionth that of the terrestrial atmosphere. Whether this atmosphere would be ionized by solar radiation to produce an ionosphere was in doubt. It was believed that the number of ions in a cubic centimeter of any lunar ionosphere would be at least 1000 times fewer than those in the ionosphere of Earth.

Since the tenuous lunar atmosphere may have been dominated by volcanism or other outgassing from the lunar interior, its analysis could provide geochemical and geophysical information about lunar evolution. Therefore the hypothetical lunar atmosphere was not treated as unimportant. Investigation of a lunar ionosphere, if present, could supplement atmospheric measurements because the ionosphere has to be derived primarily from the lunar atmosphere. Interactions of charged particles from the ionosphere with the lunar surface would reveal information about the surface materials.

A mass spectrometer deployed on the lunar surface by Apollo 17 astronauts confirmed the presence of hydrogen, helium, neon, and argon in an extremely tenuous atmosphere. But no traces of volcanic gases were found. The amount of helium corresponds to the amount expected to be collected by the Moon from the incoming solar wind. But the amount of hydrogen and neon are greater than would be expected from the solar wind source. The argon concentration is consistent with origin from the radioactive decay of potassium in lunar material.

Abundances of these gases vary between day and night as expected; argon increases at dawn, which shows that the gas condenses during the cold of the lunar night but evaporates when the Sun rises and the surface temperature increases.

Earlier Apollo missions, 12, 14, and 15, placed experiments on the surface to determine the total concentration of gas in the lunar atmosphere without trying to ascertain its constituents. The instruments were so sensitive that they could detect the aura of gas escaping from an astronaut's suit as he walked by the gauge (figure 4.12). The experiments confirmed that the pressure of the gases at the surface of the Moon is extremely low; 100 billion times less than the terrestrial atmosphere at sea level. This was, in fact, much less than had been suggested earlier. Every cubic centimeter of Earth's sea-level atmosphere has a trillion times more gas molecules than the lunar atmosphere. For all intents and purposes the Moon is, indeed, airless.

In ultraviolet spectrometer experiments made from orbit, no lunar atmospheric constituents were detected except for a short-lived cloud originating from the descent of each Lunar Module to the surface. Measurements of gas molecules along the orbit of the Command Module showed that these most probably originated from the module itself. However, it appears that the crew observed from lunar

FIGURE 4.12 Instruments placed on the surface to detect any traces of a lunar atmosphere were so sensitive that they could detect the aura of gas escaping from the suit of an astronaut passing by a detector. (*NASA*)

orbit the effects of the presence of lunar dust that extended to above the altitude of the Command Module's orbit. As the module approached sunrise, glowing streamers were seen over the terminator region. Since the lunar atmosphere is much too thin to produce such streamers they were assumed to be caused by lunar dust, which appeared to be concentrated in layers.

Horizon glows were also observed by the cameras of Surveyor unmanned spacecraft that landed on the Moon before the Apollo missions. The Soviet Lunokhod-II also detected scattering by particles from a layer about 0.5 miles (1 km) above the surface. This dust may result from meteorite impacts throwing particles of the surface into orbit around the Moon.

While the lunar atmosphere appears to derive mainly from the solar wind, it is also lost when solar radiation ionizes gas molecules that are then accelerated by the magnetic field of the solar wind and pulled away from the Moon. The lunar atmosphere could be losing mass into space in this way at the rate of 5 gm/sec. This would balance the recharging of the lunar atmosphere by the incoming solar wind and by argon coming from the Moon's interior as the result of the radioactive decay of some lunar materials. In fact, in 1991, astronomers of Boston University searched for and found a glowing 15,000 mile (24,000 km) trail of sodium atoms streaming from the Moon along the solar wind like the tail of a comet.

• THE WIND AND THE RAYS

The Moon interacts with the solar wind differently from Earth (figure 4.13). The Earth's magnetic field gives rise to a magnetosphere that acts as a barrier to the solar wind and produces a bow shock ahead of the Earth relative to the flow of the solar wind past the planet. When the Moon enters the Earth's magnetosphere once each month close to Full Moon, the magnetic environment of the Moon is dominated by the steady magnetic field of the Earth extending downstream in the solar wind. But when the Moon is on the sunward side of Earth, close to New Moon, it is outside the terrestrial magnetosphere and is dominated by solar effects.

The lunar global magnetic field is too weak to produce a shock front in the solar wind so that the wind impinges directly on the lunar surface when the Moon is not protected by Earth's magnetosphere. On the dark side of the Moon the lunar bulk

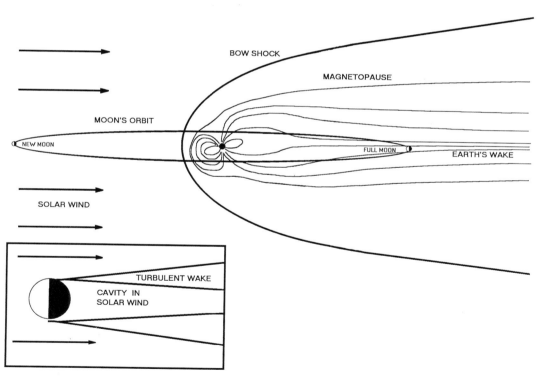

FIGURE 4.13   The magnetic field of the Earth interacts in a complex way with the solar wind, and the Moon moves in and out of Earth's magnetosphere during the course of the Moon's monthly orbit around the Earth. The Moon itself causes a cavity in the solar wind which, during part of each lunation, impinges directly on the lunar surface so that solar wind particles become incorporated into the lunar regolith.

essentially screens the solar wind and causes a cavity that is essentially empty of solar wind particles. It is the ability of the solar wind to interact directly with the bulk of the Moon that allows solar wind particles to become incorporated in the lunar regolith and allows induced fields to provide an insight into the interior of the Moon.

The Moon is also bombarded by solar and galactic cosmic rays; highly energetic particles that penetrate exposed lunar surfaces. The tracks of the particles, which are revealed by etching lunar samples with acid, can be used to determine surface history. If a rock sample shows many tracks it must have been exposed on the surface for much longer than a sample having only a few tracks. Unfortunately the tracks are present only on the surface layers of rocks, layers that are also eroded by micrometeorites.

Examination of fossil tracks produced by energetic particles (figure 4.14) provides information about the cosmic ray and solar wind compositions and fluxes. These tracks suggest that there has not been any great variations in either source over the most recent one billion years of lunar history. However, there is some evidence to suggest there may have been changes in the ratios of cosmic ray particles of different masses during this period.

Low energy atomic nuclei from solar flares produce tracks that can be identified separately because these tracks are somewhat different from the cosmic ray tracks. It was found that solar flares do not always produce particles that would be expected from the composition of the bright outer layers of the Sun (the photosphere) in which the flares appear. This may be because ions of heavy elements are accelerated in the flare differently from light elements.

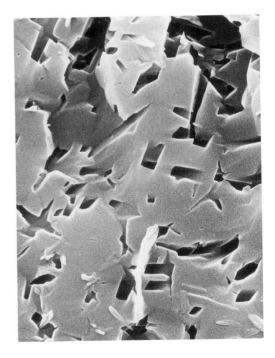

FIGURE 4.14 Cosmic rays and solar particles penetrate lunar materials close to the surface and leave tracks within them. These tracks are revealed by etching the lunar samples with acid as shown in this micrograph. The track pattern and the density of tracks provide information about the period the rocks have been exposed on the lunar surface, and also about the flux of particles and changes to it. (*NASA*)

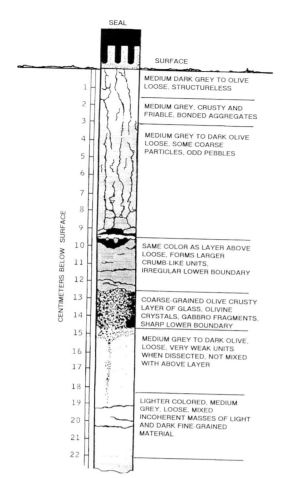

SEAL

SURFACE

MEDIUM DARK GREY TO OLIVE
LOOSE, STRUCTURELESS

MEDIUM GREY, CRUSTY AND
FRIABLE, BONDED AGGREGATES

MEDIUM GREY TO DARK OLIVE
LOOSE, SOME COARSE
PARTICLES, ODD PEBBLES

CENTIMETERS BELOW SURFACE

SAME COLOR AS LAYER ABOVE
LOOSE, FORMS LARGER
CRUMB-LIKE UNITS,
IRREGULAR LOWER BOUNDARY

COARSE-GRAINED OLIVE CRUSTY
LAYER OF GLASS, OLIVINE
CRYSTALS, GABBRO FRAGMENTS,
SHARP LOWER BOUNDARY

MEDIUM GREY TO DARK OLIVE,
LOOSE, VERY WEAK UNITS
WHEN DISSECTED, NOT MIXED
WITH ABOVE LAYER

LIGHTER COLORED, MEDIUM
GREY, LOOSE, MIXED
INCOHERENT MASSES OF LIGHT
AND DARK FINE-GRAINED
MATERIAL

FIGURE 4.15 This drawing shows part of a core sample returned by Apollo 12 astronauts. It illustrates the types of materials found at different depths below the surface. Examination of these cores provides information extending over billions of years of lunar history.

Eight feet (2.5 m) of a lunar core sample (figure 4.15) preserves the record of cosmic rays and solar radiation since the beginning of complex life on Earth. As the information continues to be deciphered, it appears that for many millions of years energetic particles from the Sun have been arriving at the Moon at a constant rate and the luminosity of the Sun has not changed. There is no evidence in the lunar samples of any changes in the intensity and frequency of solar flares, the solar wind, or the galactic cosmic radiation as far back as we can look into the history of the Moon. The solar wind has not changed for 4.5 billion years. Thus the lunar studies may indicate that the Earth's ice ages were not triggered by changes on the Sun as had been speculated.

Since the radiation appears to have been quite constant, it can, in turn, be used in a kind of depth of "suntan" technique to find out how long rocks have been exposed to the solar wind and to cosmic radiation. This technique reveals how long a rock has been on the surface, which is usually less than a few million years, and how long it has been within a few inches (10 cm) of the surface, which is usually between one and ten million years. The solar flares are used for the first measurement, and the deeper penetrating cosmic rays are used for the second. Track densities show that buried samples were once exposed on the surface but have been turned over by the lunar "gardening" process. From solar flare tracks scientists estimate how long a sample was exposed on the surface, from cosmic ray tracks how long it was within a few centimeters of the surface, and from meteor impacts how long it remained in a deep surface layer.

Cosmic rays and solar radiation also produce radionucleides by disintegrating atoms of lunar materials. The resulting particles can be detected and identified to provide supporting evidence about the history of the lunar regolith.

The lunar samples also allowed investigators to go back and take a close look at solar abundances of heavy elements and compare these with abundances on Earth, in meteorites, and in the Moon. It seems that some material must have been contributed from outside the Solar System. These heavy elements on the Moon could not have been crystallized from the solar nebula and must be from interstellar dust grains. The last injection of synthesized elements may have arrived in the Solar System nebula as part of a supernova shock front that could, indeed, have been the trigger that formed the Solar System from the primeval nebulosity.

• QUESTIONS AND ANSWERS RAISE MORE QUESTIONS

Some important questions answered by lunar exploration—mainly from Apollo manned missions—can be summarized in the following paragraphs.

What is the Moon made of?   The same chemical elements as the rest of the Solar System, but relative to Earth it is short of nitrogen, helium, hydrogen, free oxygen, and compounds of carbon; all normally gaseous on Earth. It also lacks metals such as lead, bismuth, zinc, sodium, and potassium, cadmium, cesium, and thallium, and the noble metals gold, palladium, and platinum. The basic composition thus differs not only from that of Earth but also from that of chondritic meteorites. Still, the Apollo results suggest that materials on the Moon could be mined for construction and other uses at a lunar base. But there are many new questions needing answers concerning the postulated ocean of early magma, the history of early bombardment, how the Moon's early history ties in with and affected the early history of Earth, the regolith's record of solar and cosmic activity, and the most important unresolved mystery of how, where, and precisely when the Moon formed.

How were the surface features formed?   Mainly by the impact of bodies from space and concentrated billions of years ago. These bodies, some of which continue to hit the Moon, vary from microscopic grains of dust to major masses of material up to the sizes of asteroids. The largest objects, miles across, hit the Moon more than 3.5 billion years ago when they explosively excavated the great basins that now form the relatively flat lunar plains. These plains were built up by floods of basaltic lava flowing into the basins over a period of 800 million years. Mountains on the Moon were formed by the impacts and not by crustal movements as on Earth. The continued impact of small and medium sized meteorites on the unprotected lunar surface presents a hazard to humans working on that surface. Occasionally a larger meteorite might penetrate habitation or laboratory structures. Bases and vehicles to be used on the Moon's surface must be protected in ways to reduce the probability of damage by meteorites.

Was there ever life on the Moon?   The Moon never possessed an environment suitable for life. It never had free water or oxygen. The Moon is also deficient in carbon that is an essential part of the complex carbon-rich molecules of terrestrial life forms. There is no evidence of fossil life or present life on the Moon's surface, although as mentioned in the previous chapter microbes carried to the Moon on a Surveyor spacecraft were later recovered by Apollo astronauts when they brought back parts of the spacecraft to Earth. These microbes had survived dormant for several years in the lunar environment and reproduced actively when incubated on return to Earth. Human activities on the Moon and use of its resources will not cause ecological problems there. Also, no back contamination of Earth's biosphere can take place since the lunar surface is sterile.

Is the interior like that of Earth?   No; although we do not possess much information about the interior we do know that moonquakes are relatively rare events compared with earthquakes and are less intense and deeper seated than earthquakes. Also, the Moon is solid and rigid to great depths compared with Earth. It may have a central, partially molten region, but it may not have a dense iron-rich core. Any core that it might possess would have to be small compared with that of Earth. While Earth's lithosphere is about 120 miles (200 km) thick and rests on the top of a viscous mantle consisting of partially molten silicate materials, the Moon's crust rests on top of a rigid shell that extends to a depth of at least 500 miles (800 km). The stability of the lunar surface makes it an ideal place to locate large astronomical instruments and other scientific experiments that require a vibrationless environment.

What rocks are found on the Moon?   Igneous rocks mainly, basalts in the mare region and metamorphosed or brecciated rocks on the highlands. There are no terrestrial type sedimentary rocks. Breccias are the equivalent of sedimentary rocks on Earth but they were produced as the result of 'weathering' by meteor impacts not by deposition in bodies of water. Mare lavas are similar to terrestrial lavas except that they are richer in titanium and contain no water. Highland rocks are gabbroic anorthosites consisting of pyroxene mixed with feldspar. These various rocks will provide raw materials essential for supporting a permanent human presence on the Moon. However, all samples returned so far from the Moon are from the regolith. We need to determine the nature of the underlying bedrock and of the various lava layers such as those revealed at Silver Spur.

When did the Moon originate?   The oldest lunar rocks are about 4.5 billion years old. This is very close to the time at which it is believed that the Moon formed from the solar nebula or from the material resulting from a collision of a large asteroidal body with Earth. It is consistent with inferences that the Earth and meteorites are about 4.6 billion years old. But how close is 4.5 billion years to the actual time of origin of the Moon and of the Earth/Moon system?

Where did the Moon originate?   This is still not clear. Theories include break off from the Earth, accretion in the general vicinity of the Earth, and accretion elsewhere in the Solar System with subsequent capture by Earth. However, it can be stated that the older theory that the Moon originated from the Pacific Ocean is certainly not valid. The Pacific Ocean Basin formed only hundreds of millions of years ago. The Moon formed billions of years ago. The most popular current theory of the Moon's origin is that of a major collision between Earth and another large planetary body. But there are still many adherents to alternative theories.

How did the Moon evolve after its formation?   Following the Moon's formation it seems that at least the outer part of it melted to form a magma ocean and a crust floated to and solidified at the surface. There is, however, no indisputable evidence that such a magma ocean did actually form. There is evidence that a solid crust was bombarded and modified by a wide range of bodies that might have been the tail end of an accretion process or possibly a cataclysmic event after the main period of formation of the planetary bodies of the Solar System, or another cataclysmic event after the Moon had formed at a separate period from the formation of Earth. Toward the end of the rain of incoming bodies, several very large bodies hit the Moon about 4 billion years ago and explosively excavated the mare basins. The impacts fractured the crust to great depths. Subsequent radioactive heating produced lavas that flowed along the fractures and filled the basins in a series of flows extending over millions of years.

Before the outpourings of lava filled the basins the Moon's Earth-facing hemisphere looked very different from what it does now; it most probably looked very much like the far side does today. A large, multi-ringed basin similar to Mare Orientale dominated the Earth-facing hemisphere 3.8 billion years ago, with the huge crater of Iridium excavated between the inner two mountain rings of a large basin. The Serenitatis Basin was alongside Imbrium, and closer to the East limb was the smaller basin of Crisium.

During the following 0.7 billion years the face of the Moon rapidly changed about the time the most primitive replicating microorganisms had emerged in Earth's oceans. Vast outpourings of lava, radioactively heated from the crust buried by the ejecta blankets, flooded the basins, rising high on the flanks of the surrounding mountains (figure 4.16). Smaller craters on basin floors were engulfed; even the huge Iridum crater lost half its wall to the flows and appears now as Sinus Iridum. These flows were so extensive that they spilled into adjacent low-lying areas to form Oceanus Procellarum. Serenitatis lava also overflowed and joined with lava from Nectaris to form the Mare Tranquillitatis, site of the first landing of Apollo.

Soon after the extensive lava flows were completed, activity on the Moon ceased. Except for a few large impact craters and a continued erosion of the surface by smaller meteorites, the Moon has remained essentially quiet to the present day. Thus it carries in its rocks and surface features relics from the earliest days of the Solar System.

By 3.3 billion years ago the Moon's face turned toward Earth looked almost as it does today. However, it still lacked several important young craters with their prominent ray systems: Eratosthenes, Copernicus, Kepler, Aristarchus, and Tycho. These craters were formed when late impacting bodies spread ejecta and ray systems across the lava plains. Except for these large impacts, for 3 billion years the

FIGURE 4.16 Lava flooding on flanks of mountains is shown on this photograph of the area of Triesnecker crater with the smooth expanse of Mare Vaporum extending into the far distance. Triesnecker is about 17 miles (27 km) in diameter. A well-known system of linear clefts is associated with Triesnecker. These clefts can be seen in a relatively small telescope around First Quarter and Last Quarter when the terminator stretches across this area of the Moon. (*NASA*)

Moon's face has remained virtually unchanged in all its major features, although it continued and continues to be subjected to a blizzard of smaller particles that weather its surface. A question is whether volcanics have continued in some areas until the present time.

• THE MOON, EARTH, AND PLANETS: A QUESTION OF GENESIS

The Earth and the Moon have significant differences and some similarities that are listed in table 4.1, and are compared with the other terrestrial planets. Together with Earth, Mars, Mercury, and Venus, the Moon fits into an evolutionary pattern for planets that is based on each planet's size and chemical composition. All the planets probably formed in the same way from the solar nebula or from material resulting from collisions among large accreting bodies. Whether they all formed simultaneously is still uncertain, and there are differences of opinion whether the geological time scale discovered for the Moon can be applied on an interplanetary scale to establish the ages of similar features on other planets and satellites. This would not be true if the Moon formed from debris of a collision after most of the other planets had accreted.

Undoubtedly there was some kind of terminal cataclysmic bombardment of the Moon. This has been age dated. The question is whether it occurred simultaneously with a bombardment of Mars, which formed the Hellas and Argyre basins there,

FIGURE 4.17 The surface of Mercury surrounding a big basin on that planet appears very similar to the surface around Mare Imbrium on the Moon. These photos show (top) part of the mountain ring (Lunar Alps) surrounding Imbrium Basin, and (bottom) those surrounding the Caloris Basin on Mercury. An unanswered question is whether these basins were formed at about the same time by asteroid–sized bodies in the inner Solar System. (*NASA/JPL*)

and one of Mercury, which formed the Caloris basin. Did similar huge impacts form basins on Earth and Venus? It seems most likely that they did so, but most of the evidence has been masked by subsequent episodes of vulcanism. If it is assumed that all the terrestrial planets originated at or about the same time by accretion some 4.6 billion years ago, their subsequent evolution appears to have stopped at different stages. Mercury and the Moon went through a stage of crustal genesis followed by the formation of large impact basins that were later filled by extensive lava flows. We know that this process ended more than 3 billion years ago for the Moon. It is believed that the same may be true for Mercury (figure 4.17). Yet there remains the big question why two quite dissimilar bodies internally—Mercury is much denser than the Moon, is larger, and has a core and an intrinsic magnetic field—should have the same thermal history with their heat engines shutting off at about the same stage of planetary development. Mars appears to have proceeded a stage further to the beginning of global tectonics (figure 4.18). When this process halted, we cannot

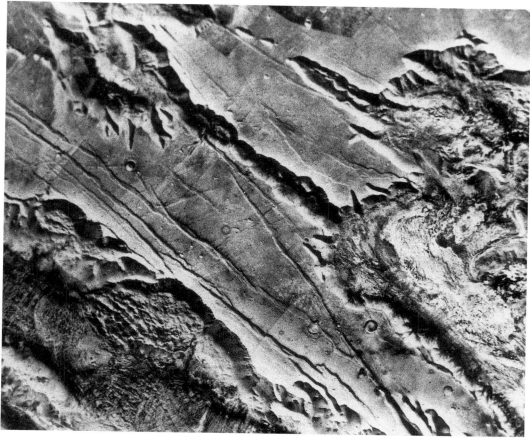

FIGURE 4.18 Canyons on Mars may indicate that plate tectonics started on that planet following an earlier episode of basin formation, but that the planetary heat engine slowed down and the tectonic crustal motions ended. Again, the big question is when did this happen. Exploration of Mars following the lunar base experience may answer this question and throw important new light on the early evolution of the Solar System. (*NASA/JPL*)

TABLE 4.1 *Moon and Earth Compared with Other Planets*

Characteristic	Earth	Moon	Mars	Venus	Mercury
Life	Yes	No	?	No	No
Water	Yes	?	As ice	No	No
Atmosphere	Yes	No	Yes	Yes	No
Impact Craters	Few	Many	Many	Some	Many
Sedimentary rocks	Water dep.	Breccias	?	?	?
Core	Iron rich	?	?	?	Yes
Mass	1	0.0123	0.11	0.83	0.037
Plate tectonics	Yes	No	Possibly	?	No
Active volcanoes	Yes	Unknown	Maybe	Yes	No
Fault zones	Many	Some	Many	Many	Some
Magnetic field	Strong	V. small	V. small	V. small	Strong
Solar distance	1.0	1.0	1.5	0.72	0.39
Formation	4.6 BY	4.6 BY	?	?	?
Geographical regions	Two	Two	Two	Two	?
Basins	Oceans	Mare	Two types	Yes	Yes
Highlands	Granite	Anorthosite	?	?	?
Basalt layering	Oceans	Basins	?	?	?
Weathering	Water	Meteorites	Dust Chemical		Meteorites

yet determine until we establish manned bases for a thorough exploration of Mars. From radar surveillance with the Magellan spacecraft it appears that Venus has developed large rifts, continents, and volcanic hot spots with large volcanoes and extensive lava flows (figure 4.19). Surface pictures from Soviet Venera spacecraft show a rock-strewn surface of Venus similar to those surfaces observed by the Viking landers on Mars.

Earth is presumed to have passed through all these earlier stages of evolution and it is still in an active stage of further development. Why Earth and Venus should be so different when so alike in size and mass is gradually emerging from the most recent exploration of Venus. The answer appears to lie not with a fundamental difference between the two planets at the time of formation but rather in their distance from the Sun, which allowed a runaway greenhouse to deprive Venus of its hydrosphere so that plate tectonics could not work in the same way they have on Earth.

To understand Earth as a planet we have to understand the other planets and why they differ from Earth. Exploration of the Moon was the first and a most important step toward this understanding. But lunar exploration was brought to an abrupt halt as the Apollo program ended and the science experiments placed on the lunar surface were turned off despite the small cost of continuing to run them and process the data. The big difficulty in comparative planetology is that of separating freakish phenomena on planets from generalities that apply to all planets of a given type. At one time, before lunar science had developed, the Moon was regarded as a freakish object, an anachronism in the Solar System. Missions to other planets have proved that this is not true. The Moon fits into a general pattern of planetary evolution. Now the major task is to try to understand the whole cosmological process that leads from primordial plasma to planets and, in at least one instance, to intelligent beings.

In this context the continued exploration of the Moon is an important stage in the rewriting of Genesis in terms of current awareness. Every civilization of note has devoted some of its energies to this important task. It would seem fitting that we who are believed to be the most advanced terrestrial civilization should do likewise with all the tools at our disposal. In rewriting Genesis we might discover ways to avoid Armageddon by self-destruction or as a result of natural causes we otherwise would fail to identify, understand, and overcome.

In common with all explorations, many new questions to greater levels of detail now reflect the deeper understanding we have obtained of the fundamental truths

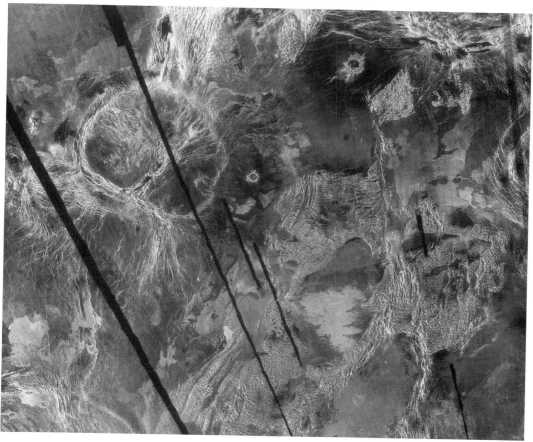

FIGURE 4.19 The surface of Venus as revealed by the planetary mapping of the Magellan spacecraft also shows evidence of widespread volcanics. This mosaic is a part of the Lada Terra highlands. It shows at top left the Eithinoha corona, a rounded, almost circular feature, which is thought to be evidence of hot magma migrating toward the planet's surface. Several lava flows and small shield volcanoes are in this structure. The bright lines extending from the corona are cracks and rifts probably produced by crustal stresses as the corona formed. The patchy mottled appearance of the plains extending over the rest of the mosaic results from lava floods. Two small impact craters are visible in the top central part of this picture. The black streaks across the image are caused by data dropouts. (*NASA/JPL*)

about the Moon. Earlier it was stressed that prior to the lunar science program many questions had been highly speculative and based on theories rather than fact. Now the questions are based on a much more solid foundation of laboratory chemical analysis and on-the-spot inspection of the lunar surface from its microscopic (figure 4.20) to its macroscopic (figure 4.21) characteristics. These questions are discussed in more detail in the next chapter.

Many unresolved questions about the Moon can be answered in part by continued study of Apollo data and samples over a long term and, more importantly, by a return to the Moon for further investigations on the surface. The coverage of the Moon so far is quite restricted. Only 20 percent of the surface has been photographed at high resolution. Geochemical coverage is only a relatively narrow swathe, as is magnetic coverage, and the polar regions are almost unexplored in any detail. For this purpose unmanned and manned missions are required, together with permanent outposts on the Moon's near and far sides. Advocates of manned and unmanned spacecraft systems need to work closely together to develop a long-range plan for the continued and sustained exploration of the Moon. This will require many compromises to keep the exploration continuing by obtaining support for it by other sections of our complex society with their different priorities and aspirations.

These questions include: Where did the Moon originate and why is this important to humankind? What was the nature of the impact flux of bodies from space? Did it tail off at one time only to be followed by a later cataclysmic event and a further separate period of impacts? How did this apply to other planets including the Earth? What is the amount of melting produced by the impacts? Could it have penetrated into the center of the Moon? Could the melting have been sufficiently widespread to appear like volcanic melt? Did a similar global melting on Earth affect the emergence of life here?

What is the underlying difference between the flooded maria and those that are not flooded? Why were the mare lava flows concentrated on the Earth-facing hemisphere of the Moon? Did Earth suffer similar global lava floods?

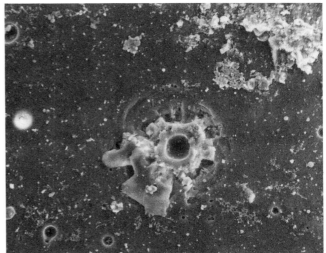

FIGURE 4.20 Based on the results of the Apollo missions we can now ask more detailed questions about the origin and evolution of the Moon when further exploration can investigate the microscopic qualities of lunar rocks. (*NASA*)

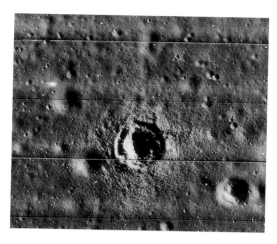

FIGURE 4.21 Also, detailed exploration of the surface, made possible by a lunar outpost followed by bases, will provide important information about the macroscopic qualities of the Moon. Cratering, as shown microscopically in figure 4.20 and macroscopically in this Lunar Orbiter image, has played an important role in lunar history and most probably in the history of the other planets and their satellites. (*NASA*)

What was the Moon like before the final stages of the impact process? What was the ancient crust that was churned by the impacts? Where did KREEP originate? Also, what is the thermal history of the Moon? Are there younger lava flows than the youngest sampled at the various Apollo landing sites? When did the Moon cease to be active? Is there still sporadic activity today?

Concerning lunar evolution we would like to know if there were ever any folded mountains on the Moon, and if plate tectonics occurred at an early stage in lunar history to be obliterated by the impacts that formed the mare basins? Was there a convective stage in lunar history? If so, when?

Other questions pertain to the lunar magnetism. What is the Moon's magnetic history? Are there strong anomalies that might offer protection from incoming energetic charged particles? Can the remanent lunar magnetism be used to outline the magnetic history of the Moon as Earth's magnetism has been used to unravel Earth's complex magnetic history?

Only a thorough exploration of the Moon can provide answers, and this exploration requires that after a program of precursor missions by orbiters and teleoperated rovers, humans spend many years exploring the Moon. In turn, this human exploration requires the establishment of permanent outposts to be followed by large bases on both the near and far sides of the Moon. Besides the great utility of Moon base experience in preparation for human expeditions and bases on other worlds such as Mars, the human long-term presence on the Moon is essential for us to understand the Earth/Moon system and its complexities. As we learn more about the Moon, we will undoubtedly be more capable of understanding that phase in Earth's history when following accretion it, too, was subjected to a furious bombardment from space as it differentiated to form a crust, oceans, and ultimately a biosphere. As stated so succinctly by G. Jeffrey Taylor and Paul D. Spudis in a paper about the inexorable intertwining of the Earth and Moon presented at the 1991 meeting of the Geological Society of America, "To understand our Earth and our place in the solar system we must go back to the Moon—with machines, with people, with gusto."

5 RETURN WITH ROBOTS

Although we have learned much from Apollo and other lunar missions we still have only rudimentary knowledge about the Moon's formation, evolution, and present internal structure. Only about 40 percent of the surface has been photomapped at high resolution, and only narrow swathes near equatorial regions (less than 20 percent of the surface) have been surveyed to find the composition of the surface materials. There have been no gravity measurements on the far side and some only along an equatorial band on the near side.

Advocates of a return to the Moon generally agree that before a lunar outpost can be planned in detail several precursor missions must be implemented. Detailed topographical and cartographic data must be obtained, much of which can be done by instrumented robotic orbiters and teleoperated surface rovers. The orbital spacecraft will be able to use the technology developed for Earth surveillance, the U.S. Landsat and military surveillance satellites, and the French Spot imaging satellite. With the end of the Cold War, it is possible that Russian surveillance technology might be used at the Moon. Samples might be collected by unmanned spacecraft and returned to Earth's laboratories for analysis. These samples will be obtained from areas of interest that were not explored by the Apollo astronauts or the Soviet unmanned surface samplers.

To select a suitable location for a human outpost or base on the lunar surface requires much more information about the characteristics of that surface and the locations and types of resources available there. Obtaining this information requires global geophysical surveys of the Moon—its gravity, magnetism, surface contours, and the like. To select a location based on what is known today would be unwise because its selection may be one of the most important decisions to be made by humanity in the near future. The lunar base will figure prominently not only in exploring the Moon, harnessing its resources, and establishing scientific observatories and research centers there, but also in developing the technology for traveling to and establishing a permanent human presence on Mars for in-depth exploration of that fascinating planet. A base incorrectly located could lead to its failure in the long run. Because of the great expense of establishing and operating a base, the financial disaster could lead to a media and consequent public negative overreaction that would end all hopes of humanity expanding from Earth into the Solar System and reaching for the stars in our time or that of our children.

Before the location for a human outpost or base is chosen it would be wise to explore the Moon with robotic unmanned vehicles for which advanced technology is now becoming commonplace in terrestrial applications. Systems are currently available or are being developed for artificial intelligence, the simulation of human thought processes by computers. Also many types of robotics are being used in manufacturing processes, and teleoperators, means of controlling complex machines remotely via communications, are being developed for a number of terrestrial purposes such as working in environments hazardous to humans. Also, experimental use of rudimentary virtual reality systems to create imaginary worlds in which humans can operate points to how humans might be connected to a powerful computer through high technology audio-visual and tactile interfaces and experience an outside world as though they were really there exploring that other world. Many of the new systems depend upon the current availability of extremely powerful and miniaturized computers, data processing systems, electronic memories, sensory instruments, and advanced communication interfaces between humans and machines.

While it seems unlikely that there will ever by a substitute for human geologists working in the field, much initial work can be done by teleoperated robots, especially work that requires generalized or repetitive tasks. Humans, however, are unexcelled in their ability during field work to take advantage of surprises and unexpected discoveries resulting from their ability to recognize patterns and make analyses synthesized from seemingly unrelated field observations of complex and subtle geological materials. Humans are without doubt essential for geological field study to obtain information at many levels of detail, but robotic machines can be designed to undertake most of the reconnaissance work to determine the broad character of the geologic features of the Moon and the processes that may have caused them.

• POST APOLLO ROBOTIC MISSIONS

Soon after the final Apollo mission NASA began studying unmanned missions to build on the Apollo results and continue lunar science through the use of unmanned spacecraft before humans would return to the Moon. In 1974, for example, NASA's Goddard Spaceflight Center looked into a Lunar Polar Orbiter mission for launching in 1980. Subsequently the study of this mission continued at the Jet Propulsion Laboratory (JPL), which had led the U.S. exploration of the Moon with its Ranger and Surveyor spacecraft. The JPL study validated mission concepts derived earlier by NASA-Goddard and showed that a spacecraft could be designed to fulfill many interplanetary missions as well as a return to the Moon. The spacecraft was called Terrestrial Bodies Orbiter (TBO). JPL personnel briefed NASA in June 1976, presenting a plan for a new start for a Terrestrial Bodies Orbiter—Lunar (TBOL) targeting for a launch to the Moon in 1980. Around this period European space enthusiasts developed a plan for a similar mission to the Moon, which they called Polar Lunar Orbiter (PLO), and the Soviets also studied a similar mission. Unfortunately for lunar science, nothing was done. The Moon remained undisturbed by visitors from Earth!

It was not until 1983 that an unmanned return to the Moon was again pushed, this time by the Solar System Exploration Committee of the NASA Advisory Council. A Lunar Geoscience Orbiter (LGO) mission was recommended for launch in 1989. Once again it was suggested as one in a new line of relatively inexpensive spacecraft designed for planetary exploration. By October of that year JPL had a Development Flight Project for a Mars Observer based on this new spacecraft concept, and this became a new start for NASA. By 1985 the concept of a return to the Moon continued to be developed to the stage at which the Lunar Geoscience Orbiter was renamed Lunar Geoscience Observer and several workshops were organized to identify its mission objectives and spacecraft requirements. But the LGO was low in a list of priorities and it was not selected as a mission to follow Mars Observer until 1986. Many studies followed to refine the Earth–Moon trajectory and lunar orbit, instrumentation to be carried by the spacecraft, and strategies to collect and return date gathered by these instruments. But by early 1987 the lunar program encountered more trouble. The nation was rapidly running into monumental fiscal troubles as it continued to squander enormous resources on a military buildup. NASA was forced to instruct JPL to reduce activities substantially in connection with the Lunar Geophysical Orbiter, and because of lack of funds the launch date for the Mars Orbiter was pushed further into the future, from 1990 to 1992.

In the usual on-and-off again political seesaw that has plagued much of the U.S. space activities, work on the LGO was authorized to continue in 1988, but the launch date had soon slipped to 1995 at the earliest. The mission was renamed Lunar Observer, yet was still not authorized as a new start.

A typical baseline mission profile for a Lunar Orbiter or a Lunar Observer is shown in figure 5.1 adapted from a Jet Propulsion Laboratory report that reviewed requirements for an orbital survey of the Moon. A typical spacecraft configuration that might be used is shown in figure 5.2, also adapted from the same report (*Lunar Observer: A Comprehensive Orbital Survey of the Moon*, Rex W. Ridenoure, Ed., JPL D-8607, April 1991).

The mission planners at JPL also studied other lunar missions in an endeavor to move the nation toward a return to the Moon by showing that a start could be

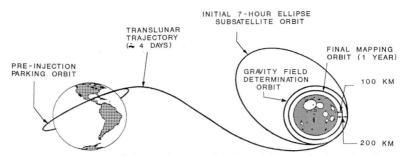

FIGURE 5.1 Mission profile of the path from Earth to Moon of a baseline Lunar Orbiter as studied by the Jet Propulsion Laboratory. (*after JPL illustration*)

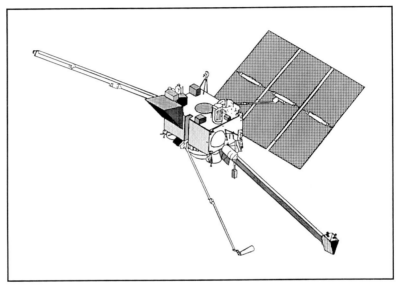

FIGURE 5.2 Possible configuration suggested for a Lunar Orbiter spacecraft. (*after JPL illustration*)

made with inexpensive spacecraft. One was for a small "Get-Away Special" lunar satellite launched from the payload bay of a space shuttle, and others were for several "bare bones" missions of very small spacecraft carrying single instruments. These were called Lunar Scouts. However, none of these simple missions have been approved and none are likely to become a reality much before the end of this century. Meanwhile a Lunar Exploration Science Working Group had been formed at NASA with the purpose of providing information for a *Moon/Mars Human Exploration Initiative*.

As the launch date for a possible Lunar Orbiter mission continued to slip into the late 1990s because a new start for this mission failed to materialize, many different studies were made of scaled-down versions of an unmanned mission to the Moon. But following the announcement of the *Space Exploration Initiative* in late 1989, a Lunar Orbiter mission became a high priority of that program. Lunar Orbiter was scheduled for a launch in 1996 and a Phase A study proposal was invited. Again the budget axe soon fell. Congress decided to withhold most of the funds for studies concerning the *Space Exploration Initiative*. Once more the launch date slipped, and this time almost out of this century; to 1999. The name of *The Space Exploration Initiative* was changed to *Mission from Planet Earth* following the release of a report on the U.S. future in space. (*Report of the Advisory Committee on the Future of the U.S. Space Program,* N. Augustine, Chairman, December 1990, Washington, D.C., Government Printing Office). Another report (*Exploring the Moon and Mars: Choices for the Nation,* U.S.Congress, Office of Technology Assessment, OTA-ISC-502, July, 1991, Washington, D.C., Government Printing Office) discussed the option of using unmanned spacecraft as a viable option of the *Space Exploration Initiative.*

The detailed report discussed many reasons for going to the Moon and the need to develop robotic explorers and what their roles would be in exploring the Moon and Mars. However, it did not provide any clearly stated conclusions but instead provided a background review of technical and mission options for the members of Congress to make decisions. The report emphasized that there were many options, and that budget constraints could seriously restrict space operations in the coming years. It also stated that a commitment to major manned exploration, rather than an economically viable mix of manned and unmanned missions to the Moon and Mars, could consume most of NASA's restricted budget to the detriment of other space science programs and the *Mission to Planet Earth,* a program to monitor the terrestrial environment from space.

• A PRIVATE ENTERPRISE PROSPECTOR

In the past few years the national space program for a human expansion into the Solar System has had no shortage of debates, studies, or exotic names, but very little positive spaceward action otherwise. As a result, scientists and spacecraft engineers were becoming extremely unhappy with the lack of firm commitments of the federal government to a clearly defined national space program. Several nongovernment activities blossomed aimed at returning to the Moon. Lockheed Missiles and Space Company, Sunnyvale, California, in a recent private enterprise project studied a Lunar Prospector spacecraft that would be used to search for water deposits. Brenda Forman, writing in the January 1991 issue of the San Francisco Branch News Letter of the American Institute of Aeronautics and Astronautics, stated; "If we want to put a base on the Moon, or if we plan to use the Moon as our jumping-off place for Mars, then we absolutely have to know what the Moon is really made of, how it's shaped, and most particularly, whether there's any water there."

The Lunar Prospector would be built with off-the-shelf hardware or with equipment donated to the project. The study was funded originally by Space Studies Institute of Princeton, New Jersey, and now is also supported by Lunar Exploration, Inc., which is a group of scientists who worked on Viking and on Apollo, together with engineers who volunteered their time and creativity to push ahead with the space program again. An aim is to show that the technology exists for private enterprise to get into the space business seriously. The estimated cost of the Lunar Prospector mission is less than $15 million, if the Russians can be persuaded to provide a free launch with one of their large fleet of available launch vehicles, as seems likely. The Prospector consortium reportedly entered into an agreement in 1990, in the form of a letter of intent, to use what was then a Soviet launch vehicle for Prospector.

At the time of writing some 20 percent of the cost of the Lunar Prospector had been promised according to Forman. A full-scale mockup of the spin-stabilized 290 pound (132 kg) Prospector has been built by Omni Systems, Inc., of El Segundo, California, and major design reviews have been completed. The Prospector would orbit the Moon for about one year at 60 miles (100 km) altitude above the surface, carrying a magnetometer, a reflectometer, an alpha particle spectrometer, and a gamma ray/neutron spectrometer. The project is on hold at the time of writing until the necessary capital for the venture can be raised. One might feel impatience that capitalism's much lauded private enterprise system in which CEOs and baseball players have annual incomes approaching the same ballpark figures, cannot raise a few million dollars to send Prospector quickly on its way to the Moon, but one should not lose sight of the fact that even NASA's government supported return to the Moon has been delayed for more than a score of years and is still not firmly committed. In fact, it could take NASA 30 years to get back to the Moon, three times as long as it took to develop Apollo and land the first humans on another world. So much for progress and technology leadership toward the end of the twentieth century!

• SOLAR SAILING AROUND THE MOON

In the early 1980s a French group, *Union pour la Promotion de la Propulsion Photonique* (U3P) proposed a solar sail race to the Moon and more recently there have been proposals to stage such a race in honor of the 500th anniversary of Columbus's first voyage to the Americas. The World Space Foundation, Pasadena, has been pushing for solar sail activities in space for many years. In fact, the French proposal was stimulated by reports about the Foundation's solar sail designs. Several international meetings followed and in August 1991 a meeting at CNES Headquarters, Paris, agreed on plans for a 1994 launch of three pioneering solar sail spacecraft in a race to the Moon. The spacecraft would represent groups of European, American, and Japanese sponsors and would be launched from either the Arianespace or the Energia launch complexes using either an Ariane or a Proton rocket launch vehicle. The three spacecraft would be launched together and each would be given 14 days to pass a start line.

The goal of the race is to perform a flyby of the Moon and send back photographs of the far side. The unpiloted spacecraft would be judged on skillful navigation with solar sails, first to the finish line (judged by return of the photographs), and efficiency of the solar sail design. The American spacecraft will carry messages designed by students as the equivalent of cave paintings to be read by future generations. Also some scientific payloads will be carried by each spacecraft and the results judged on the basis of their contributions to space science. The messages, which are referred to as SpaceArc, were gathered worldwide by The Rochester (New York) Museum and Science Center. They might be picked up centuries from now as the old World Space Foundation spacecraft, having flown by the Moon, will continue to orbit the Sun between the orbits of Earth and Mars.

All spacecraft are supported by private enterprise groups interested in furthering human expansion into space. Many sponsors are from international aerospace groups. As of early 1991 the World Space Foundation entry into the race to the Moon had received half a million dollars of monetary or material support.

So where can we look for a quick return to the Moon? Obviously not to the Russians; nor in the near future to the European space agencies, who have not participated in plans for lunar exploration for many years. There is, nevertheless, considerable interest among the European space community for a return to the Moon. The Directorate of Science of the European Space Agency (ESA) initiated a study in 1990 of the motivations for such a return to the Moon and the role that ESA might play in the undertaking. While not specifically planning for a European mission, the study was aimed at how the European space community might participate in an international venture to explore the Moon.

A Lunar Study Steering Group was formed. It included experts of many disciplines under the chairmanship of Professor Baisiger of the University of Bern and used six supporting study teams. These teams directed their efforts to establish significant benefits to be derived from a return to the Moon in the fields of general lunar science, interferometry, very low frequency astronomy, astrometry, solar physics, and life sciences.

A four-phase mission approach to lunar exploration was suggested. The program would start with a lunar orbiter polar mission to provide complete geophysical, geological, and geochemical maps of the lunar surface. Next there would be missions to place rovers and science stations on the surface. Their task would be to determine the Moon's internal structure and its chemical and mineralogical composition. The next phase would be to return samples of the lunar surface from sites considered to be most important—including those on the highlands and on the far side. The final stage would be to set up the lunar outpost on the surface for continued exploration by humans living there.

The ESA report states that in the area of science from the Moon the first priority appears to be to develop a long baseline interferometric capability that would cover a range of wavelengths ranging from the ultraviolet region of the spectrum to submillimeter. An important aim would be to obtain high-sensitivity microarcsecond imaging and astrometric observations of very high precision. Such a system would be used to observe objects in distant galaxies, in our own galaxy, and in the Solar System.

Additionally there would be major science objectives in connection with human operations on the Moon. These would include obtaining information necessary for establishing the base on the Moon and testing an artificial ecosystem maintained on a planetary body other than Earth as a prerequisite for human operations on Mars.

The results of the lunar program studies were presented to ESA's Director General for implementation sometime in the future. They will be published (1992 or 1993) as an ESA scientific publication. No schedule was established for when these recommendations might be implemented. Although the Ministers of several nations that met in Munich in November 1991 to review Europe's space activities did not approve any long-term space exploration plan, they did recommend that ESA should open its programs to broader international participation. The main elements of a long-term space plan for Europe are the piloted Columbus and Hermes space vehicle systems. Since the Ministers did not give a go-ahead for a plan, opponents of the piloted European space program felt that their viewpoint of a robotic rather than human exploration, including a return to the Moon, had been accepted. Other European officials, however, are not so sure that this is how Europe will explore space and they await a further conference of Ministers in Spain in late 1992, which might decide the issue more clearly.

• PROBING ROBOTS FROM JAPAN

Possibly the lead will be taken by Japan where a new unmanned mission to the Moon has been authorized for 1996. The Japanese plan to build on their success in sending the Muses-A spacecraft to the Moon in 1990. The main spacecraft, named Hiten after its launch from Kagoshima Space Center in January of that year, made several Earth/Moon orbits and placed a small (12 kg) satellite, Hagaromo, in an elliptical, two-day period orbit around the Moon. It is understood that the Japanese plan is for a more advanced scientific spacecraft to investigate the magnetospheric tail of Earth and dispatch several penetrator probes to the lunar surface. Such probes were proposed by the author more than thirty years ago in a technical paper entitled "Why Go To The Moon?" It was published in the international aerospace magazine *Interavia* in 1959 when I was working in the advanced programs office of an electronics company specializing in instruments for data gathering and reduction in connection with tracking and testing at a major missile range. "Capabilities of landing a payload on the Moon with a hard landing are present in current military rocket vehicles," I wrote. "One of the first and most important missions . . . would be to determine the nature of the maria. . . ." I suggested using the Titan or Atlas ICBM booster rockets equipped with high performance upper stages to send the probes to the Moon and ascertain the nature of its surface, especially if there was evidence of volcanic flows. I continued with a description of the type of instrumentation and how data would be telemetered back to Earth, before the more difficult soft landings could be obtained of the type proposed in a contemporary Rand report. That report had just been published and suggested that a 50 pound (22 kg) package of instruments might be soft landed by application of existing military hardware.

The new Japanese orbiting spacecraft, Luna-A, launched by an improved booster rocket developed by Japan, will carry three 30 pound (13 kg) probes that will be separated while the spacecraft is in an initial elliptical orbit. Each will use a small rocket motor to deorbit and descend to the lunar surface. Two penetrators will impact the Earth-facing hemisphere, and one will go down on the far side to provide a seismic net of global range. Each probe will carry a seismometer to record moonquakes, and a heat-flow instrument to establish how much heat is coming to the surface from the Moon's interior so as to verify or disprove the heat flow derived from two measurements by Apollo missions at mare borders. The data gathered, as explained elsewhere in this book concerning the Apollo program results, is important in trying to determine the internal structure of the Moon.

After release of the probes, the main spacecraft will be placed in a circular lunar polar orbit about 60 miles (100 km) above the surface to relay information from the probes back to Japan. Later Moon missions by Japan will use an improved spacecraft with more probes, followed by lunar roving vehicles possibly as precursors to a Japanese outpost on the Moon. With a movement afoot by Japan to establish at United Nations a means to end the global arms race and seek general disarmament, it could very well be that in view of the financial difficulties being experienced by the Russians and the U.S. as the result of an unprecedented arms race over the last 50 years, Japan might become the leading spacefaring nation of the next century.

• IMPORTANCE OF GLOBAL SURVEYS

A return of robots to the Moon is mandatory before humankind embarks on the long-term program to establish bases there. A Synthesis Group report—chaired by astronaut Stafford—published in 1991 (*America at the Threshold,* June 1991, The White House, Washington D.C.) investigated alternative ways to establish a lunar base and explore Mars. It discussed the importance of a global geophysical network on the Moon to gather much needed information, and teleoperated rovers to explore the surface before humans go back.

We need a comprehensive global assessment of the physical, chemical, and environmental properties of the Moon if we are to address many fundamental questions of a scientific nature and about choice of a site for a lunar outpost. To select a suitable site we require information about the surface topography, the thickness, composition, and structure of the regolith, the mineralogy of the lunar surface, the distribution of rock types, the general geochemistry, and the chemistry and properties of the lunar atmosphere, not only at the site itself but also in those parts of the lunar surface that will be accessible to exploration from the base site.

Two types of global surveys of the Moon are needed. A polar orbiting spacecraft can provide maps of all the surface for geochemical, geophysical, and mineralogical information. A global network of surface instruments can provide much needed information about the lunar interior and the lunar atmosphere.

The role of a lunar orbiter prior to the return of humans to the Moon seems clear. However, there have been complications arising from the inability of the nation's government to decide that a lunar program should start again. Originally the Lunar Observer was intended for science only, but following the announcement by the Bush Administration of the return of humans to the Moon, the objectives of the Lunar Observer were broadened to include obtaining data in support of the manned program and the establishment of a lunar outpost. Lack of enthusiastic Congressional support for a long-term space exploration program and for the Lunar Observer had by the time of writing reduced the objectives again to be mainly of a scientific nature. If the program is finally approved for a new start it is likely that the scientific objectives will include those described in the following paragraphs. However, much depends upon the size of the spacecraft that may be selected since there are options to send either a large spacecraft with a complex payload of instruments, or a series of smaller spacecraft each carrying only a few instruments.

A measurement of the spectrum of thermal emissions from the lunar surface will provide information about the mineralogy and thermal and physical properties of the lunar surface materials. At infrared wavelengths of 6 to 50 micrometers minerals can be identified and the composition of rocks can be determined from orbit. Thermal and physical properties can be determined from observations between 4.3 and 100 micrometers wavelengths. A visual and infrared mapping spectrometer can be used to identify specific mineral species based on spectral absorption in reflected sunlight and can show the spatial distribution of rock units and soil types.

High-resolution visual images obtained globally from orbit can be used to describe surface morphology and lead to a basic understanding of the geologic processes responsible for that morphology. Also the images of highest resolution can be used to find areas hazardous to robotic and human explorers. Topographic maps can be produced from stereo images, and all the photographic data can be recorded in digital form so that it can be processed in the many ways proved so successful for images of Earth's surface and those of other planets in revealing subtle details not recognizable in conventional unprocessed imagery.

Laser altimetry can supplement the photo surveys to define contours and relative heights of surface features to a vertical accuracy of several feet. The data also will allow for an exact determination of the shape of the Moon when used with orbital tracking data.

The global gravity field of the Moon can be mapped with a subsatellite and this can be combined with the topography data to find the interior density of the Moon and how it is distributed, and the distribution and location of gravity anomalies including mascons.

Magnetic field measurements will map the residual magnetic field and search for anomalies similar to that found at Reiner Gamma by the Apollo subsatellites. A search can be made for more intense anomalies that may provide some protection at the surface against cosmic rays and charged particles from the Sun.

Observations of ultraviolet radiation reflected from the lunar surface will offer means to check for changes in the lunar atmosphere and possibly help to detect the presence of volatiles at the poles and of high-titanium rock types elsewhere.

A microwave radiometer will provide information about the properties of the regolith at and below the surface. This may provide the needed data about the heat flowing from the interior of the Moon, supplementing that obtained by Apollo and that from the proposed Japanese probes. It will allow scientists to determine the thickness of the regolith on a large scale but, because of limited resolution, still not in sufficient detail for site location. This may require the use of surface rovers.

Neutral and ion mass spectrometers can be used to assess the various components, neutral and ionized, of the extremely rarefied but important lunar atmosphere. Since this atmosphere will affect many scientific uses of the Moon as a world to be explored and as a base for advanced scientific instruments to observe the cosmos and the Earth, its present nature and the effects of robotic and human operations on the Moon must be understood and monitored closely. Also, a radio astronomy receiver carried around the Moon on each orbit can be used to assess the radio environment in the Moon's vicinity and the merits of establishing radio observatories at various sites on the Moon.

• BACK TO THE SURFACE: SOFT LANDERS AND ROVERS

The next important precursor mission to the lunar surface involves soft landing of unmanned spacecraft carrying rover vehicles capable of exploring areas of the Moon under control from Earth, and of placing instruments on the Moon to form a global network of greater complexity and capability than the initial network established by penetrators. In the past we have visualized such rovers as large and complex machines. However, recent studies and experiments by the Jet Propulsion Laboratory and others have shown that planetary rovers can be of several types, including both small and miniature rovers capable of using today's highly miniaturized technology to explore the surface.

The Jet Propulsion Laboratory designed and built several prototype robotic planetary exploration vehicles. The most advanced of these, Rocky III, was demonstrated in late 1991 in the desolate Avawatz Mountains south of Death Valley, California. The demonstration showed that the small rover was able to travel over rough terrain and pick rock or soil samples with its manipulator arm. Rocks at the test site were about the same size and distribution as those at the Viking 2 site on Mars, with fairly large boulders strewn over a surface of gravel. The vehicle was also tested on a lava field in California's Mojave Desert, similar in texture to the lava flows on the Moon.

The diminutive vehicle (figure 5.3) measures 24 inches (60 cm) long by 20 inches (50 cm) wide, by 16 inches (40 cm) high, about the size of some laser printers, and weighs a mere 56 pounds (25 kg). The miniature rover is radio controlled and carries a compact video camera system so that the remote operator can direct operations visually.

Such small rovers can carry cameras for close-up looks at the surface and to scan the horizon. Onboard micro-machined sensors will test the lunar atmosphere and soil, spectrometers will gather geologic information, and seismometers will capture data about the interior of the Moon. Two classes of vehicles are being developed; micro-rovers are robotic vehicles weighing under 11 pounds (5 kg), and mini-rovers weigh up to just over 50 pounds (25 kg). Rocky III falls in the mini-rover category. Because of their small size and low weight, micro- and mini-rovers would be relatively inexpensive to launch to the Moon or Mars. "A new era of

FIGURE 5.3 Rocky III, a prototype for a robotic planetary exploration vehicle, demonstrates ability to go over rough terrain and pick soil and rock samples with a manipulator arm. The rover is radio-controlled and carries a video camera so that it can be remotely controlled and directed to geological sites. The rover was designed and built by the Jet Propulsion Laboratory. (*JPL*)

space exploration is made possible by advances in miniaturization technology and distributed communications," said Giulio Varsi, manager of JPL's Space Automation and Robotics Program. "I believe these advances will make possible less expensive missions and broader participation of people."

Robotic exploration of the Moon's surface might accordingly consist first of orbiting spacecraft followed by automated landers that would gather samples from specific areas of interest and return these samples to Earth for analysis. Such sampling would, however, have to be reserved for sites in which the geology is not complex. Investigation of complex sites will have to be deferred until humans can be operating again on the Moon.

• TELEOPERATORS AND TELEPRESENCE

While humans are essential for geologic field work, the cost of sending humans to the Moon and supporting them on that world makes it necessary to employ teleoperated robots as much as possible during initial explorations. A robotic field geologist was proposed by G. Jeffrey Taylor and Paul D. Spudis at a workshop held at the Lunar and Planetary Institute, Houston, in 1988 (G. Jeffrey Taylor, ed., 1990, *Geoscience and a Lunar Base,* NASA Conference Publication 3070, Washington, D.C.). The Teleprospector would be operated from a manned base on the Moon. It would provide stereo vision to the operator through twin high-resolution TV cameras relaying signals to a special headset which in turn would have sensors to direct the "head" and "eyes" of the robot automatically as the operator's head turned. A sensory glove system, an advanced high technology version of the gloves used today in some video games, would allow the operator to control a remote "hand" to touch and manipulate surface samples remotely as though handling them in a laboratory or in the field. So far visual reality systems, such as this would be, have not produced high-resolution images for their operators, but this is because they attempt to generate real-time video graphics of imaginary places and objects. It takes enormous computer power to generate in real time the necessary millions of pixels for a convincingly clear video display. By contrast, the teleoperator for activities on the Moon would be processing actual images and it could operate in real time at high resolution. While high resolution visual-reality video graphic systems are still in the distant future, teleoperators using images from TV cameras have already proved practical and extremely useful. Such activities have been demonstrated often in terrestrial applications such as clean-up operations within nuclear reactors and other hazardous areas, manipulation of radioactive or biologically hazardous materials in laboratories, and in deep sea salvage or exploratory operations on wrecked ships.

To inspect rocks on the Moon the geologist would direct the Teleprospector to move around a site so that rocks could be viewed from several angles before being disturbed. In geological field work it is vital to look at samples from the viewpoint of their relationship to the site and surrounding geological formations, and to record this information before moving the samples. With teleoperated rovers this requirement can be satisfied. Ideally, the machines would have to have the ability to range widely over the lunar surface and spend considerable time without maintenance at sites remote from the human base. A Teleprospector must also be able to climb over boulder fields and negotiate various slopes, which might require a rover to be equipped with legs as well as or alternative to wheels. Ideally, too, an arm similar to that of a human with tactile feedback to the human operator should be included. A sense of touch is important in field geology. Also another arm should be included that is equipped with tools to drill for cores or hammer at samples. There should be instruments to perform simple analytical tasks in the field, and storage containers so that material can be sealed within them and returned to the base in as close to pristine condition as possible.

Telepresence, which is the ability of the operator to experience being at the geological site without physically having to be there, appears to many geologists to be an essential feature of robotic exploration of the Moon. Human powers of observation and synthesis are vital to geological field work. A major question to be addressed, however, is the importance of quickly sensing what is being observed at the remote site. Signals take close to two-and-a-half seconds to travel between Earth and Moon. If an operator can be trained to accept a change in vision several seconds after moving his or her head without experiencing disorientation, then teleoperators may be controlled from Earth. Otherwise they could be used only after an outpost has been established on the Moon's surface where signal time delays would not be apparent to an operator requiring telepresence.

• PATHWAY TO THE HUMAN OUTPOST

Several artist's impressions of the first stage of utilizing robotic vehicles to assess lunar resources prior to establishment of a manned base were generated some years ago at the Jet Propulsion Laboratory (figures 5.4, 5.5, and 5.6). After an orbiting satellite has made any necessary surveys of the chemical, physical, and morphological properties of the lunar surface, automated rovers would explore the surface to seek volatiles in permanently shadowed areas of lunar craters. Later automated processing stations might be set down on the lunar surface by unpiloted landers. Automated mining rover vehicles would gather the lunar regolith and bring it to the processors. The stations would use solar power to produce metals and gases from the regolith and store them in special modules, also landed automatically. This whole activity could be remotely controlled from Earth. Later, when humans returned to the Moon, they could use the accumulated resources to construct and operate the base. There have been, of course, many such scenarios some of which

FIGURE 5.4 Some years ago the Jet Propulsion Laboratory studied future advanced projects in space, including alternative propulsion systems such as solar sails, and missions to comets, other planets, and the Moon. Several artist's concepts were produced to illustrate these missions. This conceptual drawing illustrates an automated lunar rover seeking volatiles as it uses floodlights to peer with detectors and television eyes into a permanently shadowed crater. (*NASA/JPL*)

FIGURE 5.5   The study also suggested solar powered automated processing stations of the type shown in this drawing to produce metals and gases from lunar soil. Automatic mining rovers would be used to bring the lunar materials to the processing plant. All these units would be placed on the Moon and controlled from Earth before a human crew landed. In this way the crews would be assured of adequate supplies of vital materials and resources before beginning to set up the first habitat station. (*NASA/JPL*)

are discussed further in subsequent chapters, but only the future will show which of them, if any, is practical or economically viable. Without speculations to trigger our imagination the great human adventures would never take place and our "experts" would remain closeted in their ivory towers of specialized learning.

The sooner unmanned spacecraft can be sent to the Moon to accomplish these tasks, the sooner we will be prepared for a return of humans to the Moon and the establishment of bases there for in–depth exploration of our neighbor world. Despite the success of the Apollo program and twenty years of study, the rich and diverse geological history of the Moon cannot be read in detail. This is not surprising since geologists have been studying Earth's geology for about 200 years and still do not fully understand our planet. Nevertheless, because we have obtained samples of lunar rocks, we understand the Moon more than any other body of the complex worlds of our Solar System. Continued exploration of the Moon will help us understand the processes that brought the bodies of the Solar System to the stage they are today; the inhospitable, searing, dense atmosphere of Earth's errant twin, Venus, with its towering active volcanoes and planetwide lava flows and chasms, the cold desiccated Mars, and the life supporting Earth. Why do they differ so greatly?

Surely seeking an answer to why we are able to be here, and how long we will be able to remain, on Earth is the most pressing task of our time. It surely is of higher priority than our endless clashes of ideologies, and our fascination with playing god in attempting to stem the inevitable passage toward extinction which for eons has been the lot of the myriad experimental species of life on the dynamic and ever changing Earth. For if we do not find an answer we might inadvertently make our own planet also uninhabitable to all higher forms of life including ourselves. As a result, a cosmic experiment to develop a spreading consciousness from the Solar System may end abruptly.

FIGURE 5.6 The third step in this sequence of establishing a lunar base would be landing of human crews. This illustration shows an inflatable habitat alongside a landing module with several automated processing plants in the distance. (*NASA/JPL*)

The Moon appears to many as the best place in the Solar System to study the evolutionary processes that affected the larger silicate bodies of the Solar System of which the Earth is an example. The Moon is the most easily accessible, it holds a record of the history of the Solar System, it may be intimately connected with the origin of Earth, it probably holds evidence of episodes that led to the extinction of many life forms on Earth, its surface contains a record of the Sun's activity over eons, and it may contain a record of cosmic events that triggered the emergence of life on Earth. Additionally, as discussed in a subsequent chapter, the Moon has the potential of offering an ideal base to locate advanced instruments to look out at the universe and enable us to push our mind probes to the far frontiers of space and time. Finally, from a purely material viewpoint, the Moon may be a significant source of raw materials and other resources for many human activities around this planet and during the eventual human expansion to other worlds.

In April 1988 NASA announced that it had selected nine Space Engineering Research Centers based at universities to further the research needed for a long-range space program. The centers were established to conduct research and to develop technologies that are required to set up and operate bases on the Moon, to conduct unmanned and manned operations on Mars, and to support other space flight operations. Additionally the centers are training young men and women capable of meeting the challenge of continuing space exploration.

A center at the University of Arizona concentrates on how local planetary resources can be utilized. It conducts research to determine ways by which human habitation on the Moon and on other planets and their satellites can be made affordable, especially through use of resources found on extraterrestrial bodies. Current emphasis is on methods of producing oxygen and hydrogen and rocket propellants, use of planetary resources to produce building materials, and how these materials might be produced with minimum energy consumption. The center is also developing artificially intelligent control and communication systems for remote processing plants operating in extraterrestrial environments, and is searching for near-Earth asteroids that may be sources of useful materials to support a long-range space program.

The University has also conducted several annual symposia on the utilization of planetary resources. In February 1992, for example, a symposium was devoted to the technology of using lunar materials, including the characteristics of the lunar materials, management of energy for their extraction and use, processing of lunar materials, controlling the environment, applications of automated robots for construction and maintenance, and communications on the Moon. At the beginning of the symposium, Aaron Cohen, Director of NASA's Johnson Space Center, emphasized the technical spinoffs that might come from establishing operations on other worlds. "Through exploration of the Moon and utilization of its bountiful resources, lunar explorers of the 21st century will reap a rich harvest of new discoveries, new technologies, new processes, and new sources of energy and raw materials that will be used not just for the benefit of the lunar operation itself, but for the benefit of all humankind."

Cohen said these benefits will include "the development of large-scale renewable energy resources that could have a profound impact on Earth's economy and ecosystems by the mid-21st century; the production of oxygen from lunar materials that could greatly offset the complexity and cost of space operations, and the [extraterrestrial] mining of a variety of raw materials that ultimately could help make the lunar operation self-sufficient."

The center at the University of Colorado concentrates on space construction, dealing with fabricating large systems in orbit and the technology required to support establishing a manned base on the Moon. Studies and research encompass the materials, environment, construction equipment, and construction methods that will be involved, taking into account the constraints imposed by available launch vehicles and space transportation systems. How lunar soil can be used to protect humans on the Moon, and how robots can be used for heavy or dangerous work are also being studied.

Massachusetts Institute of Technology has a center devoted to space engineering. Its long-range goal is to develop technology that will allow the operation of complex scientific instruments, such as astronomical observatories, on a space platform. Particularly important is understanding and correcting disturbances produced by docking of space vehicles with a space station and vibrations from other activities within an orbiting space complex.

North Carolina State University and North Carolina Agricultural and Technical State University have a center that is concerned with research about missions to Mars. Areas of research currently include hypersonic aerodynamics and the use of the atmosphere of Earth and Mars to slow interplanetary spacecraft and to enter or change orbits. (This research also applies to return of lunar spacecraft into low Earth orbit.) Research is conducted on composites to develop materials that are light in weight yet strong enough for construction of interplanetary spacecraft. Also the research is aimed at developing ways to control vehicles traveling at hypersonic speeds in planetary atmospheres. The requirements for a mission to Mars and how such a mission can be designed are being analyzed. Such research is important to the lunar program because a human presence on the Moon with propellant production there could be necessary to establish an economically viable permanent human presence on Mars. The lunar and Mars programs may turn out to be highly dependent on each other.

Rensselaer Polytechnic Institute deals with intelligent robotic systems. The center at this institute is engaged in developing the theory of intelligent control systems and their algorithms for autonomous systems, and is developing laboratory demonstrations of these systems. A typical example being researched is assembly in space of a large structure of trusses by robotic machines. A mobile robot facility is researching how collisions can be avoided and risks minimized on rovers designed to explore the surfaces of other worlds.

The center at the University of Idaho is working on the design of very large scale integrated systems. Research is directed at applications such as data communication and compression, navigation, control, robotics, and processing of images. Systems are being designed in which thousands of transistors are incorporated in a single chip smaller than a postage stamp. Also, research is investigating how these integrated circuits can withstand severe radiation environments arising from exposure to cosmic rays and energetic particles from solar flares.

At the University of Michigan the accent is on space terahertz technology. The center is investigating a region of the electromagnetic spectrum between that of visible light and the shortest radio waves—a region that is important to understanding the amount of specific chemicals in the Earth's upper atmosphere—and some astrophysical phenomena. Investigating the region of the spectrum requires use of derivatives of radio and optical technologies and new semiconductors and manufacturing techniques. The center is currently working on sensor technology, integrated-circuit receivers, low-cost guiding structures to replace the common optical fibers and coaxial cables which cannot be used at the terahertz frequencies, and monolithic antenna arrays grown on single chips.

The University of Cincinnati has a center specializing in sensor, control, and management technologies to achieve high reliability in space systems. The two missions of this center are to train students and professionals in space engineering principles and to develop specific expertise and technologies in systems and component "health" management that can be transferred to government and industry. Sensors that monitor space propulsion systems and systems to correct faults are studied and researched. Techniques for diagnosing conditions of space systems, characterizing flows within systems, and researching how materials might degrade during space operations are being studied and analyzed.

Pennsylvania State University has a center specializing in propulsion engineering. Two main goals have been established; to provide a focused research effort on space propulsion engineering, and to interest students in following space propulsion careers. Research covers hydrogen management in materials for high-pressure hydrogen/oxygen engines, ignition and combustion of metallized propellants, and computational fluid dynamic analysis of flow fields in the combustion chamber and nozzle of rocket engines.

Equally as important as the technical research, each of these centers plays a major role in educating a new generation of space professionals able to step beyond the traditional technologies developed during the first half century of the space age. Through this program these children of the space age, some born since the orbiting of the first Earth satellites in the late 1950s, and some born after men first walked on the Moon, will be given the requisite skills and knowledge to open the Solar System to humankind in the 21st century.

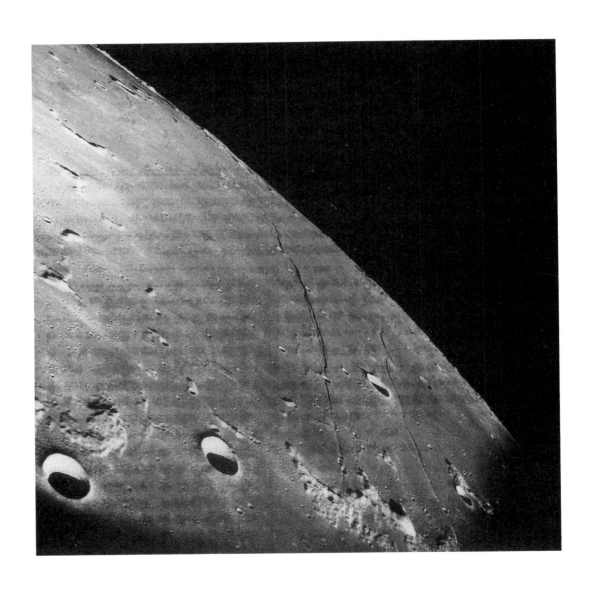

6 THE LUNAR OUTPOST

The global exploration of the Moon as a world cannot take place until lunar explorers, including scientists of various disciplines, are able to stay on the Moon for several months at least, much as humans have stayed at bases in the Antarctic to begin to understand that continent, and as the Atlantic-crossing Europeans had to establish permanent bases in the New World to understand and develop the American continent. It would not have been feasible to obtain the enormous amounts of information about Antarctica if we had sent only instrumented machines, using as an excuse that conditions were too harsh and it cost too much to establish an environment in which inquisitive and curious humans could operate on the Antarctic continent. Nor could the vast potential of the Americas have been unleashed if humans had not been willing to risk their lives exploring and developing the alien continent against natural resistance from sometimes hostile native inhabitants.

• GEOSCIENCE AT A LUNAR OUTPOST

Detailed plans for geoscience on the Moon have been outlined at a number of scientific workshops, particularly since the Apollo program. A recent such workshop was held at the Lunar and Planetary Institute, Houston, Texas, in August 1988. The proceedings were published as a NASA document (G. Jeffrey Taylor and

Paul D. Spudis, eds., 1990, *Geoscience and a Lunar Base; A Comprehensive Plan for Lunar Exploration,* NASA Conference Publication 3070, Washington D.C.).

The report points out that geoscience studies can begin even during construction of a manned base, when geologist astronauts would be able to travel several kilometers around the construction site for periods longer than was possible during the Apollo missions. Excavation equipment needed for construction of the initial outpost can also be used to dig deep trenches into the regolith for the study of stratigraphy down to bedrock, where the interface between the regolith and the bedrock can be studied. Examination of the layers of regolith is important also to obtaining a record of solar activities for up to 4 billion years. The amount of entrapped volatiles can be ascertained for the various levels. Later, as the outpost is developed, trenches would be excavated in a suitable geometry to investigate the relationships between pyroclastic layers and regolith layers seeking for correlations and checking for evidence of cyclic episodes of layering.

A variety of surface samples will be collected around the base site to provide information not only about the site itself but also about the surrounding region. Bedrock outcrops on crater walls can be studied and field investigations at various geologically interesting sites will be less time limited than they were with the Apollo missions. Sites will also be revisited if needed after an initial examination of discoveries—a common practice during terrestrial field work which was impossible with Apollo-type missions.

Once the base is operational and field work can occur for several miles around it, several important geological missions seem appropriate. One would be to sample pyroclastic glasses that are known to be unmodified samples of magma that carry volatiles entrapped in gas bubbles and provide information about the amounts and composition of those volatiles in the Moon's interior. Geologists operating from the base would make comprehensive maps of deposits on the lunar surface; their composition of minerals and the color, size, and shapes of the particles of the regolith. Differences in pyroclastic deposits will be looked for.

Another field task would be to map geological units below the surface by use of an electromagnetic instrument that induces responses from shallow depths, and active seismometry using explosives and thumpers. This exploration of subsurface units would locate highland rocks that may be buried beneath shallow lava flows and lava tube caves which may be ideal locations for some activities of an advanced base.

An extensive seismic network will be established, ideally with eight stations spaced over the whole surface of the Moon. While an initial seismic network might have been set up by penetrators, an advanced and more sophisticated set of instruments will require mobility over the surface with a range of at least 3000 miles (5000 km), and equipment to drill into the surface and locate the instruments on suitable foundations at precisely selected positions on the surface. This network, which will provide low-resolution seismic information on a global basis, will be supplemented by local seismic arrays close to the lunar base. The arrays would not only provide good information about natural and human-generated seismic activity produced by construction work, spacecraft landings and takeoffs, and intentional explosions, but also monitor the seismic environment for nearby astronomical observatories. Similar highly sensitive arrays would also be established at other locations away from the main base, especially near to scientific equipment installations that require a seismically inactive location if they are to operate effectively.

While much gravity data may be obtained from orbit, geologists on the Moon's surface can use gravimeters to investigate mascons and other anomalies, as another way of locating underground lava tubes, and possibly to search for lunar resources buried beneath the regolith.

Magnetic anomalies can be investigated by gathering samples at the location of the anomaly for transport to a laboratory at the base where they can be analyzed for their paleomagnetism. Surface magnetometer measurements of the magnetic field generally would be correlated with those made from orbit.

An ion mass spectrometer and a neutral gas mass spectrometer will be useful instruments to position at several locations around a base to monitor the lunar atmosphere for environmental as well as general scientific purposes. Their measurements would be compared with those made by a similar pair of instruments deposited in the base area before the work of constructing the base started.

When the base is completely in operation, additional more sophisticated geoscience investigations can be made with a complex of teleoperated robotic machines and human geologists. The robots would conduct much routine geological work, especially at locations where humans might find it difficult or hazardous to work. Humans would examine bizarre features which might be enigmatic to machines. There are many puzzling or intriguing features on the Moon that require sustained field work there. Examples include an unusual morphology of the floor of Murchison (southeast of Sinus Aestuum), a possible young caldera in Ina (near Haemus Mountains), the thorium-rich rocks of Archimedes (Mare Imbrium) with unusual spectral features, the volcanic features of Tobias Mayer (near Carpathian Mountains) and similar rilles, and the wrinkle ridges of the lunar plains. The report cited above lists 29 intriguing areas for geoscience investigation along a traverse across parts of the Mare Imbrium and Oceanus Procellarum. It also recommends that a capability be established at the proposed base for safely storing lunar samples and analyzing them in a laboratory facility at the base instead of returning them to Earth for analysis. The laboratory should have instruments to classify rock types, identify unusual samples and soils rich in gases, study paleomagnetism in samples, and identify samples that are unusual in subtle ways.

• SOME EARLY IDEAS FOR OUTPOSTS

So what are likely scenarios for establishing an outpost, followed by a base or bases on the Moon? How might we use lunar materials and develop the Moon as a resource for many human activities and a beginning of human expansion into the Solar System? Many scenarios have been suggested and studied to various levels of detail for at least 50 years. Only a few of these can be selected for discussion in this and the following chapter. The author apologizes to the originators of those he has not had space to discuss in this book.

In 1962 at a meeting of the American Rocket Society in Los Angeles, a technical paper by C. William Henderson described a Douglas Aircraft Company study for a lunar survival shelter that could provide living conditions for lunar explorers beyond the capabilities of Apollo's Lunar Module.

A cylindrical shelter containing a robot vehicle would be launched from Earth, using a Saturn V class vehicle, and soft landed on the Moon. The robot vehicle, controlled from Earth, would emerge from the shelter through a nose door (figure 6.1) and would emplace a portable nuclear reactor to supply power for life support and communications. Next the robot would scoop up lunar soil and cover the shelter to protect it from small meteoroids, radiation, and temperature extremes.

The environment within the shelter would be controlled from Earth and thoroughly checked before the arrival of a crew in a separate Apollo-type vehicle. Additional structures would be shipped from Earth in subsequent launches and landed near to the base. Crew members would match panels, attach pipe fittings, and make electrical connections to the various components of a base that had been designed for easy assembly by men in spacesuits.

FIGURE 6.1   A study by Douglas Missiles and Space Division in 1962 showed how remotely controlled vehicles could be carried to the Moon to prepare a habitat for a crew that would arrive later. This drawing by Douglas artist Ron Simpson shows a remotely controlled robot vehicle towing a nuclear power source from an unmanned lander. This vehicle was equipped to scoop up lunar soil later and spray it over the module to protect it from micrometeorites and radiation. The module would become the habitat for the human crew. (*McDonnell Douglas*)

At the same meeting Bruce M. Carr of Gallery Chemical Co. described how lunar materials might be used to provide essential consumables by hydrogen reduction of iron oxides and silicates expected to be present in the lunar regolith. The process might pass pulverized regolith through an electric arc—the electricity coming from a nuclear generator—in a stream of hydrogen. Alternatively the rock might be compressed to form an electrode for the arc. One extremely innovative approach suggested was to avoid processing lunar materials in bulk. Carr proposed using an underground hydrogen arc jet similar to a constricted arc cutting tool. The arc jet would be used to dig a hole in the lunar regolith and penetrate to a depth that could contain a pressure of a few atmospheres (one atmosphere is equal to terrestrial surface pressure at sea level). The top of the hole would be sealed, with the arc jet remaining in the cavity while hydrogen would circulate from above and through the arc. A processing unit on the surface would recover water and oxygen. Such a unit weighing less than 10 tons could produce five pounds of oxygen per hour, Carr said.

The 1960s saw many such proposals for lunar bases and for utilizing lunar resources, mainly in the United States and Soviet Russia. The Soviet concepts were revealed for the first time at a meeting of the International Astronautical Federation in Warsaw in 1964. A paper by B. V. Zayonchkovskiy, L. N. Lavrenov, and B. G. Tarasevich explored some of the architectural problems of establishing a base on the Moon. Basically the paper looked at the various ideas for providing habitable quarters, including an air-supported structure placed in an excavation and covered with lunar soil for protection. The proposed design had an airlock on the surface accessed through a vertical tunnel from the buried habitation.

About this period IIT Research Institute published the results of a design study also suggesting pneumatically erected structures for a lunar habitat. These were lightweight units designed to act as temporary habitats for astronauts during the building of a permanent base. The structures would be transported to the Moon in a small package. Unloaded by astronauts, each unit would be rib-supported, as opposed to purely pneumatic, and would provide working space for two astronauts and their life support. Several units could be assembled into a star pattern with airlocks connected to provide accommodation for a larger crew of lunar explorers. A wall consisting of foam within inner and outer membranes would be expected to dissipate much of the energy of small meteoroids. Actual punctures would be repairable if easily located. The problem would be intense solar flares, which might require that one module be buried and that astronauts be crowded into it to take shelter during a solar storm.

Hughes Aircraft Company also studied the fabrication of foamed polyurethane structures in a space environment. One fabricated in 1962 was a 10-foot (3-m) diameter balloon inflated and rigidized in a vacuum. Its walls were one to two inches thick. The material melted at 180°F and foamed in a vacuum such as that on the Moon without needing a separate blowing agent.

For extraterrestrial bases new fabrication methods will be encouraged and the development of capabilities to form and erect structures on the Moon probably will have important spinoffs back to terrestrial applications, as happened during the development of the Apollo program. Thrusting technology into new environments of exploration has always led to important new benefits for humankind.

A versatile lunar base system must be capable of supporting several scientific missions under a wide range of lunar conditions. Early studies concentrated on an expandable modular base system that could be adaptable both to small outposts and to larger permanent installations housing up to 18 people. In 1963 NASA instituted studies to consider the general concept of a complete base system and three major elements of such a base—a life-support system, a nuclear power plant, and a regenerative fuel system for surface vehicles. The Office of the Chief of Engineers, U.S. Army administered the studies of the power plant and the fuel system.

Additionally NASA made studies to define the research and development effort required to provide a lunar research capability and the experimental facilities required here on Earth to support such a research effort, and to establish a schedule and cost estimates for a lunar construction research program. Results of earlier studies by the Air Force and the Army (project Horizon) were utilized. Areas of research included characteristics of the lunar soil, soil movement and excavation techniques, construction materials, structural design, power generation, storing and handling of life-supporting atmosphere, water supply and sanitation, construction tools, and human engineering and training. The relative advantages of installing the lunar facility on or below the surface were discussed as possible alternatives.

While in the early 1960s the exact course of post-Apollo lunar exploration could not be predicted, many people associated with the program fully expected that many probable future lunar missions would require support by facilities on the lunar surface. It was accepted that while the size, lifetime, and operational purpose of such facilities would vary with specific post-Apollo missions, all would have to perform the basic functions of providing shelter, life-support, power, communications, and surface mobility.

It was also acknowledged that any acceptable concept for a lunar base had to capitalize on the similarities that would be certain to exist in any lunar facility yet still retain sufficient flexibility to accommodate many uncertainties. Flexibility appeared to be readily achievable if some departure from an optimum design was acceptable.

In general it was assumed that a lunar base would consist of a family of prefabricated modules assembled on or below the surface in a variety of arrays suitable to different mission objectives. All the modules had to be transportable in lunar logistic carriers. At that time the carrier available was the Saturn V, which would have restricted modules to a diameter of less than 20 feet (6.1 m) and shaped as a 20-foot diameter, 8-feet (2.4-m) high cylinder topped by a right circular cone with a height of just over 20 feet. The weight of the module had to be less than 25,000 pounds (11,340 kg). Each module had to be capable of being stored unattended on the surface of the Moon for one year. To support missions of long duration, the modules had to be capable of supporting a crew for up to six months before requiring the return of personnel to Earth. Also, supply from Earth was estimated to be restricted to one Saturn V every eight weeks.

There were many subsequent studies for establishing a base on the Moon. Several early ones concluded that the Saturn V could be used to carry to the lunar surface all the materials and modules needed to construct a relatively inexpensive semi-permanent outpost. That outpost would have been built up by a series of compartments each of which could be fitted into the payload capacity and shape restrictions of the Saturn V booster configuration mentioned earlier. An initial simple shelter would have been expanded by delivery of more modules later. A report of the President's Space Task Group in 1969 anticipated the serious hiatus in exploration of the Moon that would follow the series of Apollo landings and would last from 1975 through 1981. However, it did not expect the inordinate and inexcusable inactivity lasting to the end of the century.

NASA engineers stated at the time that derivatives of the Apollo system could be used with little additional expenditure for development. Modifications to the adapter section of Saturn would have consisted of sealing the upper part so that it could become a manned laboratory and living quarters for a crew of two. The lower part would not have been pressurized but would have been stocked with supplies for the crew sufficient for them to live on the Moon for just over 100 days. The lower part would have contained also an adaptation of Apollo's Lunar Module (LM) engines to permit a soft landing of the adapter section on the lunar surface. If, additionally, the surface engine of the Command Module were used as well as the LM engine for the descent to the lunar surface, an even larger payload of 12 tons could have been landed on the Moon. To do this two Saturn V launches would have been required; one to carry a normal Apollo system to take two crew members into lunar orbit, where they would have transferred to a Lunar Excursion Module, leaving the Command/Service Module unattended in lunar orbit, the other to carry a Lunar Base Module. In lunar orbit the Base Module would have been attached to the Service Module and its engine then used to decelerate the module sufficiently for its own engine to achieve a safe landing after the Service Module had been detached from the Lunar Base Module. The two men in their Lunar Excursion Module would then have descended to the lunar surface to land close to the Base Module. Then they would have been able to transfer to the Base Module with some of their equipment.

After about 100 days on the Moon the crew would have moved back into the Lunar Excursion Module and used its ascent stage to carry them back into lunar orbit, where they would have transferred to the Command/Service Module still orbiting there. With this module they would next have returned to Earth in the usual Apollo fashion, splashing down in an ocean.

If this or a similar program had been funded and carried out, American astronauts could have been at a permanent outpost on the Moon by 1972. Although the mission would not have been as cost effective as a lunar base mission designed from the ground up, and based on a space transportation system such as the originally envisioned heavy lift space shuttle, it would have provided experience for the design of a more cost effective system when a shuttle with sufficient lifting capacity became available.

However, the nation decided to abandon the technology developed by Apollo and retreat from any expansion into space. Instead, it concentrated on the reduced-capacity space transportation system for operations in LEO without any long-term congressionally approved plan for the exploration of the Solar System, although such a plan had been defined by NASA officials, various commissions, and private enterprises. Even unmanned exploration was seriously curtailed with no new starts.

There are many possible arrangements for a lunar base arising from the essential flexible, building-block approach that seems mandatory both for the development of a base from simple outposts and the use of lunar facilities for many purposes of exploration, science, and further human expansion into the Solar System, and the use of lunar resources in Earth-orbit and elsewhere, possibly for lunar manufacturing, and space-based solar-power systems.

The simplest configuration is that of a single module used as an outpost similar to the use of the Lunar Excursion Module during the Apollo missions, but for a longer period of human activity on the Moon, ideally measured in weeks or months instead of the days of the Apollo missions. The next logical development will be to use several modules: a central life-support module, and modules for communications, power, fuel, and maintenance. An even larger base for extended missions on the Moon would add more modules, such as regenerative life-support equipment, additional communication capabilities, regenerative fuels production, and maintenance support for the added systems.

A fully fledged lunar base would have still more modules: to extract life-support materials from lunar resources, extract fuel from lunar resources, and provide additional maintenance support for these new systems. Also it would add more complex science modules and more sophisticated vehicles and materials-processing equipment with supporting modules for both. For such a base there might be a complex of five to ten life-support modules housing a score or more of active personnel of different engineering, operations, and science disciplines. This advanced base will be needed to explore and exploit the lunar resource, making available a new economic commons for Earth's peoples.

The basic module of a lunar outpost is the personnel shelter, a small, temporary outpost that would be equipped with its own integral life-support, power, and communications equipment, and would be capable of housing several people. When the shelter is expanded into part of a permanent base, these integral systems would be used as emergency backups should the separate life-support, power, and communications systems fail.

A central life-support system is essential for all but the simplest temporary outpost. This system will be modular in design; the number of modules will depend upon the size of the lunar base and its human contingent and the degree of autonomy provided for separate units of the base. To begin with, the modules will provide only those functions absolutely necessary for the survival of a few personnel. As the facilities on the Moon become permanent and house more people, the life-support system will be developed to include regenerative systems and recycling to reduce the need for resupplying consumables from Earth. Ultimately means will be developed to generate many life-support materials from resources on the Moon available in the regolith or through deep-mining operations.

Electrical power can be generated using solar cells, but only for relatively small installations. For a permanent large base on the Moon use of nuclear-generated electrical power will be essential. This is particularly so because of the vulnerability of solar collectors to environmental conditions on the lunar surface, particularly the insidious nature of lunar dust.

Fuel for vehicles, equipment, and mobile power sources to be used for mining, road making, and establishment of scientific operations as well as exploration will most probably be chemical. Initially the fuel must be shipped from Earth in fuel storage modules. This will be extremely expensive; transporting each pound of freight from Earth to Moon may exceed $10,000 using conventional rocket propelled systems. But, writing in the June 1987 issue of *Aerospace America,* Graeme Aston of the Jet Propulsion Laboratory showed there are alternatives. He described how a fleet of four ion-propelled lunar transports could ferry 20 tons of payload every hundred days to a lunar base. The ion propulsion engines would derive their power from solar arrays. Such ion engines have been under development for some time, principally at NASA Lewis Research Center. It is possible that several decades in the future the cost of transporting materials from Earth to Moon might be reduced considerably by innovative new propulsion systems.

For early operations on the Moon, however, fuel might have to be regenerated from engine exhausts using energy from nuclear power plants. Fuel components also might be generated from lunar resources. Although oxygen is readily available from lunar materials, obtaining hydrogen on the Moon is a problem. Recent radar surveys of Mercury have suggested that water may be trapped in the polar regions of that hot planet. If this is a true interpretation of the radar results, it seems likely that the Moon may also have been able to trap in its polar regions water and other volatiles from comets that must have impacted the Moon since its formation. If water ice is discovered at the lunar poles, hydrogen could be produced by electrolysis; however, the water itself may be a much too valuable resource for it to be used as a source of hydrogen. Indeed, recycling of water from engine exhaust gases may be mandatory. One of the great beneficial fallouts from bases on the Moon might well be the efficient recycling systems that must be developed for humans to survive on the Moon. These will find applications here on Earth where they might otherwise never be developed because the need is not so pressing as it will be on the Moon.

Communications and control modules will provide local communications among lunar facilities and vehicles, communications between Moon and Earth, and navigation aids to spacecraft and to vehicles on the lunar surface. Also there may be need for Earth-based monitoring and control of lunar facilities to reduce requirements for personnel on the Moon and to maintain modules and outposts on the lunar surface that may be on standby only, without human personnel, for varying periods.

Vehicles and mobile equipment for use on the Moon (figure 6.2) will be used for transferring crews and freight among landing sites, base modules, and outposts, for materials handling, surface modification such as road making and landing site grading, transportation, reconnaissance, and exploration. Some vehicles will be specialized; others should be multipurpose in their design.

Modules of several sizes will be needed to maintain equipment, vehicles, and science experiments. Specialized modules also will be needed for science experiments and for long-term science activities such as construction shacks for erecting and logistics units for operating radio telescopes and other astronomical observatories.

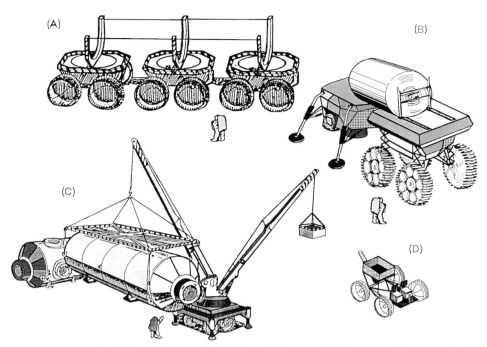

FIGURE 6.2 Several different types of vehicles will be needed for operations on the Moon. These various types have been suggested in NASA-Johnson studies. Top left is a train configuration of adaptable vehicles to handle transport of cylindrical modules. Individual vehicles can be converted as needed for other uses. Top right is a vehicle designed for unloading of cylindrical modules from a freight lander. Bottom left is a mobile crane used to emplace modules at the habitat site, and bottom right is a small two-man truck. The human figures in spacesuits give an idea of the scale. (*After NASA reports*)

• UTILIZING LUNAR RESOURCES

NASA has organized several workshops to study the post Apollo activities on the Moon and to evaluate potential sites of lunar resources. There are exotic materials that if present on the Moon would considerably assist obtaining funding for a return. Foremost is water, followed by other volatiles such as carbon, hydrogen, nitrogen, noble gases, chlorine, fluorine, and sulfur. Also zinc, cadmium, indium, mercury, lead, germanium, and halogens are concentrated into the regolith by heating from solar radiation and meteorite impacts.

Apollo missions found that within lunar rocks water is present in minuscule amounts only; it probably originated from the interaction of the oxygen in lunar rocks with hydrogen from the solar wind. There are three possibilities for finding hydrogen, which is, of course, an important rocket propellant. These are hydrogen implanted in the regolith from the solar wind, hydrogen in water ice that may be present in shadowed areas near the lunar poles, and hydrogen that may be trapped below the surface in volcanic areas.

Recovering hydrogen of solar wind origin will demand the processing of extremely large bulks of lunar soil. Probably only a few grams can be recovered from each ton of regolith processed by heating. Nevertheless, this process might be used if the processing is also yielding other useful materials as well as hydrogen.

As yet we do not know if there is polar ice on the Moon. Rovers such as the Lunar Prospector mentioned in the previous chapter should be able to answer this important question. A polar orbiter with a gamma ray spectrometer and neutron detectors could be used for preliminary surveys for initial detection, which would then be supplemented by a surface rover to ascertain exact locations and to determine recovery conditions. Ice-laden rocks may not, however, be on the surface, but buried beneath the surface regolith that has been churned by meteorite impacts at the polar regions as it has elsewhere on the Moon. The rovers may have to make surveys on site using seismic and electromagnetic soundings to find out how deep any ice-bearing rocks may extend.

The presence of hydrogen, or water, in volcanic rocks beneath the surface is purely speculative. However, there are areas on which features and colorations suggest that volcanic activity may still be occurring on the Moon. There are strange collapsed depressions, dark halo craters, unusual surface cracks, surface brightenings, and obscurations, most of which are located near the edges of the mare basins where volcanism might be expected. These need to be investigated in detail by surface rovers drilling deep cores and possibly later by human missions.

Areas of interest include, for example, the area of the Gruithuisan volcanic domes on the border of Mare Imbrium south of Heracleides Promontory on Sinus Iridum (figure 6.3), which is believed to be one of several volcanic sites where rocks may contain useful mineral ores that will be needed to construct and develop the human outpost.

In general lunar volcanic sites are hard to identify from Earth by telescopic observation. Terrestrial-size volcanoes are not easy to recognize among the chaotic jumble of impact craters and debris. There is nothing on the Moon like Olympus Mons on Mars, or the other mighty Martian and Venusian volcanoes. Fortunately the flat areas of the maria make searching easier. Extremely fluid lavas flowed across the mare surfaces, and there are many examples on these surfaces of the sinuous rilles that were the conduits of lava streams. The lunar volcanoes are not like Mount St. Helens, or Vesuvius, or Mount Fuji, but rather like the rounded domes we see around the Kilauea caldera on the island of Hawaii. They appear as rounded blisters on the lunar surface rather than high peaks. They are a few hundred feet in height and several miles in diameter, so they can only be seen from Earth at very low sun angles; that is, when they are very close to the morning or evening terminator, the boundary between day and night on the Moon.

Such volcanic domes are found around the edges of the mare basins. While many do not show any volcanic pit at the summit, others do exhibit a small rimless summit crater. Areas of the Moon where these volcanic domes are known to exist are in the region of the crater Cauchy (figure 6.4) on Mare Tranquillitatus near Proclus, an area near Arago on the opposite edge of Tranquillitatis, near the crater Keis between Mare Humorum and Mare Nubium (figure 6.5), near Hortensius just

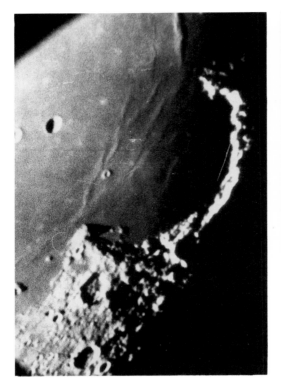

FIGURE 6.3 There are areas on the Moon where volcanic features may indicate the presence of volatiles. (a) One such area is the Gruithuisan volcanic domes south of the Heracleides Promontory of the Sinus Iridum. This bay in Mare Imbrium is shown at sunrise and sunset on these photos by Gilbert Fielder taken at Pic Du Midi Observatory. Gruithuisan is just off the field of view on the right side of the pictures. (b) Another fascinating area is on the border of Oceanus Procellarum near the lunar equator at about 60 degrees west longitude. The large feature in the west-central part of this photograph is Reiner Gamma, with a number of volcanic domes toward the distant horizon. (*NASA*)

FIGURE 6.4   This photograph looks across Mare Tranquillitatis and shows the Gauchy rille and scarp. The crater Gauchy lies between the two. Two volcanic domes are to the left of the crater beyond the scarp. (*NASA*)

west of the prominent crater Copernicus, and near the Marius Hills in Oceanus Procellarum (figure 6.6).

The area around the bright crater Aristarchus (figure 6.7), where astronomers have frequently reported seeing brightening and obscuration of lunar features, is also extremely varied geologically. It contains different types of craters, fresh and flooded, a profusion of intricate rilles, and volcanic features. The fascinating Schroeter's Valley (figure 6.8) originates in a volcanic dome that has been called the Cobra's Head. The valley has an internal valley on its floor indicating different stages of lava flows.

Other precursor missions will obtain data important to selecting the location where the outpost will be established, based on providing access to many different

FIGURE 6.5 The rilles on this photograph border Mare Humorum. The crater at the top left of the photo close to where the three rilles emerge from the mountains is Campanus. To its left and almost at the edge of the picture is a group of small hills the leftmost of which is a prominent volcanic dome known as Keis Pi. (*Photo by author*)

geological features, general scientific usefulness, access to and from Earth, and protection from the various hazards of the lunar environment.

Several precursor missions probably will require human crews who will expand on the type of exploratory work accomplished by the Apollo astronauts. Missions to the polar regions of the Moon might reveal that ice is present there in the lunar regolith or at levels accessible beneath it. Such missions might be required if lunar polar orbiters carrying gamma–ray and neutron spectrometers detect ice at depth in polar craters whose floors are always screened from solar heating. If ice is found by exploration at these polar locations, either on the surface or more likely within cores extracted by drilling through the regolith, the astronauts would be required to explore the areas further to check on the feasibility of establishing outposts there to erect and subsequently service water extraction plants.

• STEPS TO THE OUTPOST

The profile of the first phase of the lunar exploration is shown in table 6.1, which is adapted from NASA's *Report of the 90-day Study on Human Exploration of the Moon and Mars,* November 1989. It assumes the operation of space station Freedom or some equivalent established by a European/Japanese space consortium, by the

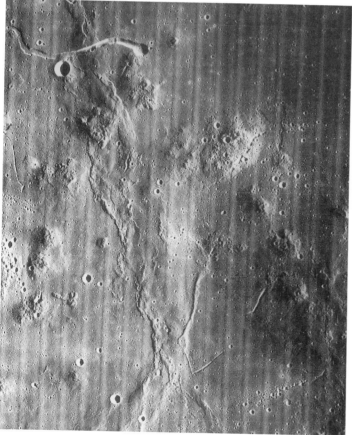

FIGURE 6.6 The Marius Hills region includes many volcanic domes and is targeted as an important area for future exploration in search of lunar volatiles and other resources. (a) An oblique view of the region with the crater Marius close to the horizon. (b) A near vertical view of part of the region showing the complexity of volcanic features, domes, cones, flows, and collapsed lava tubes. (*NASA photos*)

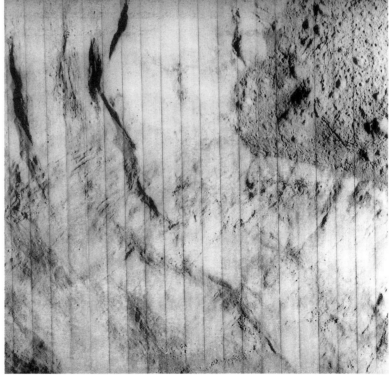

FIGURE 6.7 Another interesting region with a multitude of volcanic features is that surrounding the young crater Aristarchus and its more ancient companion Herodotus from the surrounding highlands of which emerges the Schroeter Valley. Aristarchus is the brightest feature on the Moon. The inner crater walls have slumped terraces with lava flows. By contrast, Herodotus has a lava filled flat floor. (a) A general view of the region, Aristarchus on the left. (b) Part of the interior wall of Aristarchus and the crater's jumbled floor. (*NASA Photos*)

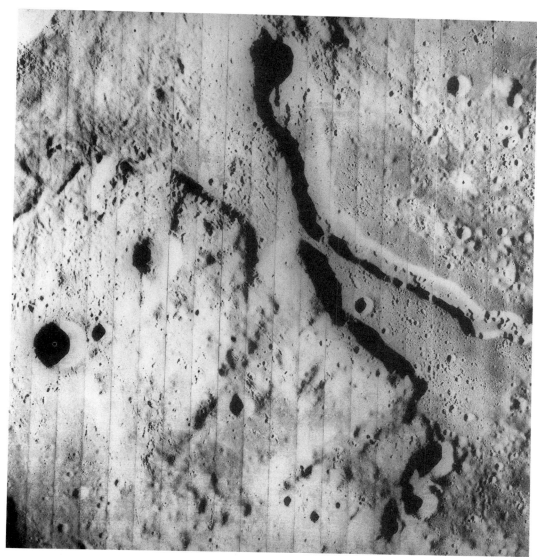

FIGURE 6.8 The area of the Cobra Head where the Schroeter Valley emerges from the highland mass surrounding Herodotus. The Valley has a sinuous channel running along its flat floor. It was formed most likely by a second episode of lava flow after the main flow that eroded the big valley. (*NASA*)

Russian commonwealth nations, or possibly by China, the development of a heavy-lift launch vehicle to move large payloads to the orbit of the space station from Earth's surface, and development of spacecraft to transfer between low Earth orbit (LEO) and low lunar orbit (LLO), and between LLO and the lunar surface.

The ways in which the various components of the space station will be developed and applied are shown in the table. For example, the truss structure would be expanded for the lunar operations, and the power and thermal system would be increased from an output of 125 watts to 175 watts for operating the reusable lunar

TABLE 6.1 *First Phases of Human Exploration*

Freedom Unit	Verification LTV Flight	Expendable LTV Operations	Reusable LTV Operations	Lunar Operations
Modules	U.S., Japan, Europe + Habitat	U.S., Japan, Europe + 2 Habitats	Same	Same
Truss Structure	Trans. Boom Lower Keels & Lower Boom	Same	Same	Trans. Boom Lower & Upper Keels & Boom
Power & Thermal	125 kwatts	125 kwatts	175 kwatts	175 kwatts
Crew	10 permanent	10 permanent 4 transient to Moon	12 permanent 4 transient to Moon	12 permanent 4 transient to Moon
Vehicle Processing	Service track Assembly for Lunar vehicle	Enclosed LTV Hanger	Same	Enclosed LTV Hanger
Remote Manipulator	One	One	Two	Two
Mobile Transporter	One	One	One	Two

transfer vehicles and for supporting operations at the Moon. The crew, originally consisting of only 10 permanent members at Freedom, would be increased to 12 permanent and 4 transient members for beginning the lunar outpost. A remote manipulator and a mobile transporter used at Freedom would be increased to two of each to support the outpost operations.

It is important to accept that there are many ways to ensure human expansion into space. For example, during the summer of 1991, a joint U.S.–Soviet private study of missions to Mars held at Stanford University showed that human missions beyond Earth can be accomplished at a much reduced cost compared with those estimated by NASA for U.S. involvement alone. The study concentrated on the base on Mars and concluded that it could be established at a cost of $60 billion over a period of 20 years; that is, only $3 billion per year (one-third of the current NASA annual budget), which is well within the space budgets of the U.S. and the Soviets. The program would use the Soviet Energia space booster. An international team could even do it alone without U.S. involvement if necessary. Such international private enterprise activity might also be developed in the form of a consortium to establish bases on the Moon, much as innovative commercially oriented ventures moved Europeans to India and the Americas following initial explorations. It is important to be aware that other nations see the importance of being a space-faring nation. Japan, for example, is making rapid progress into the space age with a new launch and research complex in Tanegashima, an island about 100 miles (160 km) south of Nagasaki and one of four Japanese launch complexes. A large construction company is making studies of long-term space programs including the establishment of bases on the Moon, while other Japanese space experts contemplate how they can get into the space station business if the U.S. Freedom project falls through.

Currently, Japan is developing large rocket engines together with vehicles to launch communication satellites, and there are plans for a mini space shuttle, at first robotic and then later manned, that can provide a transportation system for crews and cargoes between Earth's surface and LEO. Looking beyond the mini shuttle, Japanese engineers are conducting feasibility studies of aerospace planes to service space stations, the space stations themselves, and the lunar bases that they will support. The technology is being developed for Japan to move aggressively into space as it has with technology expansion on the Earth. Awaited is a government decision that the time is ripe, and the financial resources should be allocated, for Japan to be included among the leading spacefaring nations.

A non-American Moon or Mars project could be developed from the proposed international cooperation in what was to have been the Soviet 1994 Mars unmanned missions and the Soviet space station experience. The Soviets even as late as mid-1991 were still expanding their space station capabilities by additions to Mir, despite serious general problems of the Soviet economy and political unrest. Also when Europe finally becomes economically united and brings into the economic community some of the eastern European nations, possibly including Russia, their large aerospace industry, no longer able to contribute weapon systems against those nations that formerly made up the Warsaw Pact, will undoubtedly turn to the space initiative. Already there have been several proposals to develop single-stage to orbit vehicles that will make important contributions to establishing space stations as stepping stones to bases on the Moon and Mars.

The Stanford study is important concerning the Lunar Outpost because it shows that a rethinking of how we go about moving humanity into space and retreating from national competition into a global cooperation can achieve human expansion into space most economically. Establishing outposts and bases on the Moon could draw upon the lift capability of Energia and the space station nodal capability of Mir. So while the Soviets did not compete in the race to land men on the Moon, they certainly led the world in space station technology and in ensuring human operations for long periods in the space environment.

There are, of course, several possible scenarios for establishing a permanent human presence on the Moon. If humanity decides that this expansion into the Solar System should begin at this stage in the development of our species, there will undoubtedly be many much more detailed studies to decide which scenario shall be followed. As mentioned in the previous chapter, such studies will require input from robotic missions in order for them to be optimized in terms of economics, and scientific, social, and political returns for the enormous investment in human effort. But we should not lose sight of other equally expensive human investments such as the era of building the magnificent cathedrals of Europe. Those earlier activities dedicated to uplifting the human spirit despite much poverty and physical hardships of daily life were considered extremely worthwhile by the civilizations involved, and they required generations of human effort for their completion.

The comments that follow are concerning one of several currently suggested scenarios for a return to the Moon. After new lunar transportation node space stations have been placed in LEO by heavy-lift vehicles or a revived Saturn V launch vehicle program, or assembled there from modules carried to LEO with the space shuttle transportation system or its developments, or Energia vehicles supplied by the Russians, three more launches of a heavy-lift vehicle will provide sufficient capacity to carry a lunar payload, four crew members, and transportation and propellants to the space station. There the crew, payloads, and propellants are placed aboard the lunar transfer spacecraft that will carry them into a low orbit around the Moon. In that orbit is a lunar descent vehicle that at the beginning of the program will be sent from Earth orbit ahead of the crew. During extended lunar operations this vehicle will be traveling between low lunar orbit and the lunar surface, and arriving crews will meet the vehicle that has ascended to the lunar orbit from the lunar surface.

Crew, payload, and propellants are transferred into the lunar ascent/descent vehicle, which then carries them to the landing site on the lunar surface. The lunar transfer spacecraft then returns to the space station in Earth orbit to be refurbished for another mission. Should crew members be rotating in their tour of duty they will be carried back to the space station during this transfer.

The transfer spacecraft will be maintained at the space station, the lunar landing/ascent vehicle will be maintained at the lunar base landing site. Each crew will be able to remain at the lunar outpost for up to one year.

The lunar missions to support the base there will be of two main types: crew transfer and freight transfer. A piloted mission carries four crew members and some cargo from LEO to the Moon and returns a crew of four and a limited cargo from the Moon to LEO. Cargo missions carry freight only to the Moon, and the spacecraft are either discarded and perhaps used on the Moon, or are returned empty to LEO. Both types of mission will use common vehicles, except that crew carriers use a habitable cab while cargo carriers use a payload pallet.

The establishment of the base on the Moon begins with a cargo mission that not only tests the operational capability of the system but also delivers the first modules of the base to the landing site on the lunar surface. It delivers an unpressurized rover vehicle that can be manned or remotely controlled, equipment to prepare the site for the lunar outpost, and equipment necessary to offload cargo from the ascent/descent vehicle. The rover vehicle will be controlled from Earth to explore the immediate vicinity of the landing site to seek an appropriate site for the main base.

A second flight, also without a crew, will carry the module for the initial habitation: living volume, airlock, power system, and environmental control and life-support system. This module will be placed on the surface by remote control and then covered with lunar soil to protect and insulate it.

The third flight carries a crew of four who will remain on the Moon for 30 days. They will check out the habitation module and all its associated support systems, use the rover to survey the vicinity, and set up some preliminary science experiments.

Subsequently additional cargo flights and manned flights will develop the outpost to a permanent occupation status during which crews may be rotated back to Earth as required. When this configuration has been proved to be operationally successful, additional flights will provide for a much larger habitation module, construction shack, maintenance modules, and additional power capabilities. Eventually crews will be able to stay on the Moon for as long as 600 days during which they will practice for the later manned mission to establish a permanent human presence on Mars. A manned pressurized roving vehicle (figure 6.9) will permit a wide range of

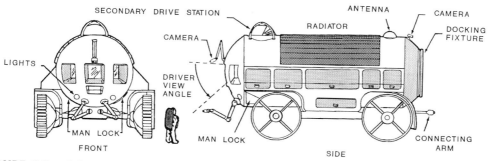

FIGURE 6.9 Advanced personnel transportation systems will be required as well as simple rovers such as that used by Apollo. This design suggested in a NASA-Johnson study would provide accommodation for several crew members in a shirtsleeves environment and would be able to travel many miles over the lunar surface. (*After NASA diagram*)

explorations over the lunar surface, and complex scientific instruments, such as radio astronomy arrays, will be set up.

• ROLE OF ORBITAL NODES: STATIONS IN SPACE

The key to all this development on another world seems to be developing the technology of operating a transportation node in low Earth orbit. Nevertheless, there have been some recent studies that conclude that direct flights from Earth's surface to lunar orbit, similar to the Apollo missions, would be more cost effective, at least for establishing the initial lunar outpost. The U.S. version of the transportation node key is space station Freedom, which is an important project not merely for science but even more importantly for technology. In many ways it parallels the efforts of the shipbuilders who made the *Beagle* possible not for the science of Darwin, even though it did become important later, but for exploration and the development of global resources. Although the science that resulted from Darwin's having been included in the crew was in many ways spectacular, the technology of exploration was the really big benefit; it created a capability of spreading civilization worldwide and developing many new lands. Should there be no scientific returns from a U.S. space station, and this is most unlikely, the importance of the technology to be provided from space station Freedom cannot be overemphasized. It is vitally important that this program be continued if Americans are to lead the international human expansion into the Solar System, first with bases on the Moon and then onward to Mars and to the mining of the asteroids.

Space stations are by no means new ideas in the spaceflight community. They have been long recognized as an essential part of developing capabilities to operate in space on a large scale. Claims that the space station project should be canceled because it does not produce much science and starves science projects of funds are utter nonsense, idle echoes of many of the science complaints leveled years ago against the Apollo project.

When the Bostonian author Edward Everett Hale wrote a short story entitled "The Brick Moon" for the *Atlantic Monthly* in 1879, it is doubtful that anyone could then have imagined the rapid development that would carry men to the Moon within a hundred years. Hale's fiction was an account of men, women, and children accompanied by poultry being carried accidentally into Earth orbit. In subsequent years the ideas of stations in space were discussed by Kurd Lasswitz (Germany, 1897), Konstantin E. Tsiolkovsky (Russia, 1907) and Hermann Oberth (Germany, 1922) as stepping stones for journeys into space.

In 1928 the scene shifted to Austria when Captain Potocnik of the Imperial Army wrote a book, *Das Problem der Befahrung des Weltraums* (The Problem of Space Travel), which was published in Germany under his pen name Hermann Noordung. He described a manned space station that would consist of three separate units connected by cables for power and hoses for air. The crew would use a wheel-shaped module that rotated to simulate partial gravity. The other two modules were to generate power and to provide a science observatory. However, Noordung suggested placing his space station complex in geosynchronous orbit, which apart from requiring a much greater booster capability would have restricted observations to somewhat less than one terrestrial hemisphere. About this time another Austrian, Guido von Pirquet, proposed the use of a space station in low Earth orbit (LEO) to assemble missions to other planets.

The idea of a space station languished for 15 years until, in 1945, Arthur C. Clarke, writing in the British magazine *Wireless World,* suggested that a manned satellite in geosynchronous orbit would be much better for relaying television programs worldwide than would the use of high flying airplanes, as had been proposed by Westinghouse and the Glen L. Martin Co. in the U.S. Three years later, in 1948, Harry E. Ross presented a paper before the British Interplanetary Society that discussed in great detail such a manned space station, but in LEO (Harry E. Ross, "Orbital Bases," *Journal B.I.S.* Vol. 8, No.1, January 1949).

The second International Astronautical Congress was held in London in 1951 (figure 6.10); its technical theme was artificial satellites. While some papers discussed unmanned satellites, I was excited by several papers that concentrated on manned space stations, and I remember discussing the relative merits of these ideas with Eugen Sanger (figure 6.10b) who had been engaged in important work on large rocket engines and an antipodal semi-ballistic bomber during World War II.

Wernher von Braun, who was by that time in the U.S., sent a paper that emphasized the importance of satellite stations as refueling nodes for interplanetary flight. Von Pirquet discussed how a space station might be established, and Hermann H. Koelle, who would later work on the Saturn program, presented a detailed analysis of design problems associated with manned space stations. He suggested a wheel-shaped modular structure with the modules being the spent propellant tanks of the upper stages of the freight rockets. Other papers looked into construction problems, use of the station as an astronomical observatory, types of supply rockets for the establishment of the station and movement of crew members between Earth and the station, protection of the station from meteorites, and some biological problems expected and how they might be overcome. Leslie R. Shepherd, then technical director of the British Interplanetary Society that was host to the congress, summed up the importance of space stations as nodes for human

FIGURE 6.10 A serious study of manned space stations for use as interplanetary refueling nodes was the technical theme of the second international astronautical congress held in London in 1951. (a) Prominent members of the British Interplanetary Society which hosted the congress are shown here. Back row, left to right; A. V. Cleaver, R. A. Smith, L. R. Shepherd, (E. Sawyer of Pacific Rocket Society) J. Humphries, A. Slater, G. V. E. Thompson, E. Burgess. Seated, left to right; K. W. Gatland, Walter Gillings, A. C. Clarke, L. J. Carter, H. E. Ross. (b) The author and K. W. Gatland discuss proposals for space stations with Eugen Sanger and Irene Sanger-Bredt at the congress. It was a paper by Eugen Sanger about a rocket-powered aircraft in December 1934 that brought together the original group of GALCIT experimenters under Theodore von Karman.

expansion into space: "We may say that the exploitation of the possibilities inherent in the satellite refuelling station gives us a clear basis for proceeding with the development of interplanetary flight."

A few years later an even more detailed proposal for a manned satellite than that presented by Koelle and von Braun's papers was published by von Braun first in Germany (*Das Marsprojekt,* Frankfurt: Umschau Verlag, 1952) and a short while later in the United States (*The Mars Project,* Urbana: University of Illinois Press, 1953). It was a plan to construct in space, from prefabricated parts carried into orbit by winged space shuttles, a rotating wheel-shaped space station in orbit just above Earth's atmosphere (figure 6.11). This would have carried a crew complement of 80 persons. This ambitious undertaking was proposed initially for the Mars manned flights favored by von Braun, but it was pointed out that the station would have many other uses in monitoring the terrestrial environment and in communications and general scientific research. The idea of space stations had finally returned across the Atlantic; no longer a brick moon but now a technical feasibility.

In April 1958, Krafft Ehricke, then with Convair (Astronautics) Division of General Dynamics Corporation, proposed a derivation of the Atlas ICBM booster that was then in pilot production for the ballistic missile program. The Atlas manned space station was visualized as the first unit of a complete Atlas orbital system. Besides the space station, the system would have included a cargo ship and a four-passenger ferry vehicle. The fully equipped space station (figure 6.12) would have been 106 feet (32 m) long and would have weighed about 15,000 pounds (6,800 kg). Living quarters in Ehricke's design were housed in an inflatable insulated capsule in the forward section of the cylindrical Atlas stainless steel tankage. The crew capsule was to be divided into four levels, as shown in the illustration. The procedure was to deliver the modified Atlas propellant tank into orbit and then

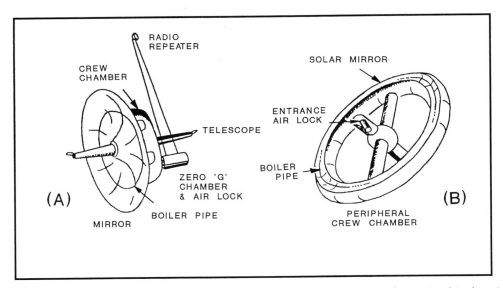

FIGURE 6.11 Two early proposals for manned space stations are shown in this drawing: the Smith-Ross station of 1949, and the von Braun wheel-shaped station of early 1950s.

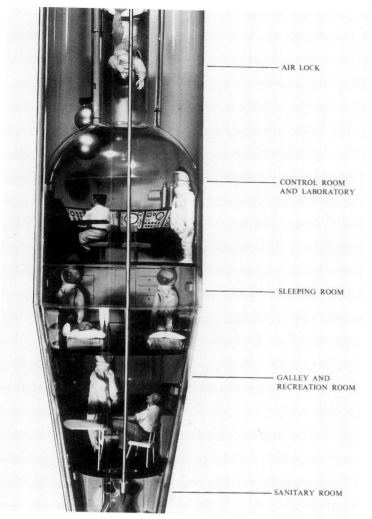

AIR LOCK

CONTROL ROOM
AND LABORATORY

SLEEPING ROOM

GALLEY AND
RECREATION ROOM

SANITARY ROOM

FIGURE 6.12 In 1958 Krafft Ehricke proposed a derivation of the
Convair Atlas ICBM booster as a
manned station for a four-man
crew. This photograph of a model
shows the internal arrangement of
the station. (*Convair/Astronautics*)

send up a cargo vehicle followed by another launching for the four-man crew for
initial setup of the station. Subsequently other cargo vehicles and crew-carrying
vehicles would have been sent into orbit to bring the station to operational status.
Finally a cargo vehicle would carry a nuclear power plant to be attached to the
station. It was anticipated that this station could have been operating within five
years of a go ahead. Outpost II, as it was named, was visualized by Ehricke as an
intermediate step to exploration of the Moon and the planets. A year later other
engineers at Convair proposed use of recoverable winged vehicles to support the
station and other space missions.

But it was not until the mid-1960s that the Department of Defense spent $1.5
billion toward an estimated $2 billion program to develop a manned orbiting
laboratory (MOL) before the project was canceled in 1969. The MOL program
officially began in 1965 when it was contracted to McDonnell Douglas Astronautics
Company, and the first manned flight to set up the station in space was planned for
1969. The 72-foot (22-m) long space station would have weighed 15 tons and been

launched into orbit by a Titan III segmented solid-propellant booster. It was to carry a crew of two for a mission of 30 days in a shirtsleeves environment.

From 1966 onward there were many changes to the design, often arising from budget cutbacks and the usual bureaucratic and political delays that inevitably pushed the proposed launch schedule back to 1972 and increased the estimated total cost to $3 billion. With about 50 percent of the cost of the program already used by 1969, there were questions about the desirability of abandoning the program, and other more serious questions about whether the program should have been started in the first instance or whether it was merely an attempt by the Air Force to participate in the manned space program. Was this a repetition of an earlier cancellation of the Air Force's Dynasoar program, which would have carried aircrews into space in a winged vehicle? That ambitious military manned space program was canceled in the early 1960s.

The Air Force then appeared to be out of the manned space business. But what about NASA? In April 1963 the Agency had invited aerospace companies to propose a study for a manned orbital research laboratory (MORL) capable of sustaining a crew of four in space for one year. In December of that year NASA awarded a contract to McDonnell Douglas to refine the DOD manned orbital laboratory concept on which it was working so that it would meet the NASA objectives. However, in 1965 a joint NASA-DOD agreement removed NASA from the space station program and it was agreed that the MOL of DOD would continue to be developed as the U.S. space station. But, as mentioned above, the MOL was subsequently canceled.

NASA started afresh. A classified preproposal conference took place at NASA's Marshall Space Flight Center at Huntsville. Contractors were invited to submit proposals to establish a space station using the complex of launch vehicles, facilities, and spacecraft available from the Apollo program. This Apollo Application Program (AAP) was intended to demonstrate and expand human capabilities to operate in space, to conduct observations of the Earth from orbit, and to prepare crews for long distance space missions.

By January 1967 the AAP program had been defined. The orbiting second stage of an uprated Saturn I booster was to be converted into a habitable module when its propellants had been consumed. An airlock would have been incorporated in the tank through which astronauts, carried into orbit separately by a Saturn, would have entered the module. This workshop would have provided an economical manned shelter for humans to orbit Earth for long periods. A later version of the mission was to use a larger Saturn IV-B second stage, which, by using the powerful Saturn V launch configuration, could be sent into orbit without any propellants having to be carried within it initially. The big dry tank could thus be fully equipped as an orbiting laboratory before launch. Three crew members transferring from Earth to the laboratory and later returning to Earth would have used the Apollo spacecraft, a contract to modify four of which was awarded to North American Rockwell in February 1969. This whole plan was very similar to that which the Soviets used to set up their initial space stations.

In 1970 I stood inside a full-scale mockup of a space station designed by North American Rockwell ready for the planned launch of such a station in 1977. Inside a 33-foot diameter cylinder towering 70 feet high were decks containing bunks, staterooms, control centers, galleys, and wardrooms. There were washrooms and storage spaces and environment conditioning sections, together with research laboratories. Access to the six decks was through a central staircase. The plan was that this space station core would be launched atop a Saturn V rocket. When in orbit the cylinder would be attached by cable to the final stage of Saturn and the whole set rotating with the Saturn third stage at one end of the cable and the space station module at the other to simulate gravity in the space station. At this time in 1970, the big question was whether or not the planned space shuttle would be operational in time to ferry crews to and from the station.

Meanwhile McDonnell Douglas had also been working on a design for the space station paralleling the study at North American Rockwell. Both studies had shown that a big space station was technically feasible. However, then as now, the budget for the station was inadequate, and maintenance in orbit was regarded as prohibitive in cost. So the accent began to change from a large station to a modular design that could be expanded later when funds became available. Over twenty years these attitudes have not appeared to change. The same problems face space station Freedom. The question is whether the nation is prepared to commit sufficient resources to become a serious participant on the undoubted future of stations in space and further expansion into the Solar System. As President Kennedy emphasized concerning Apollo: "If we are to go only halfway, or reduce our sights in the face of difficulty . . . it would be better not to go at all." Having spent enormous amounts of money and human creativity to open space for human exploration it would, however, seem a great pity that the U.S. should now throw away this investment in the future and relinquish the opening of the Solar System treasure chest to other nations.

The first American space station was scheduled to be in orbit by 1972, and beyond that interim station NASA looked forward to a much larger station with a crew of up to 50 members by 1980. Contracts were awarded to McDonnell Douglas and to North American Rockwell to study the larger space station project and develop plans for such a station. The result was Skylab, an interim space station to test out the new technology.

Skylab was intended as an experimental space station based on many proposals that had been developed over the previous decades. It was a 110-foot-high complex of highly versatile laboratories (figure 6.13) able to perform many important space medicine experiments under conditions of microgravity and some important science experiments in space, including observations of the Sun and of other celestial objects, and to find out how effectively humans could perform in the space environment for relatively much longer periods than the lunar missions had required.

In the two years since I had walked through the North American Rockwell mockup at Downey to my tour through Skylab at McDonnell Douglas, Huntington Beach, there had been great progress in designing amenities for the astronauts who would spend weeks in the experimental U.S. space station. In September 1972 the huge Skylab orbital workshop was handed over to NASA for its journey via the Panama Canal to the Florida launch site for mating with the Saturn V launch rocket. Casper W. Weinberger, then director of the Office of Management and Budget, and James Fletcher, NASA Administrator, who were present at the ceremony, stressed that the Nixon Administration was committed to a sustained

FIGURE 6.13 Skylab was the first U.S. orbiting laboratory, a precursor to space station Freedom. (a) This cutaway artwork shows the interior arrangement of the two-level crew compartment within a modified Saturn S-IVB rocket structure. One level contained workroom, sleep compartment, wardroom, and waste management compartment. The other level was a storage and experiments area. Solar array panels generated electrical power for the orbital laboratory. The Command/Service Module of the Apollo system is attached to the workshop. The Apollo Telescope Mount with its own solar arrays is attached to a docking adapter connecting the workshop to the Command/Service Module. (McDonnell Douglas) (b) A TV picture relayed from Skylab during a mission shows astronaut Gerald P. Carr preparing to use a treadmill-like exercise device. One of the purposes of the Skylab program was to determine the effects of prolonged spaceflight on humans. (*NASA*)

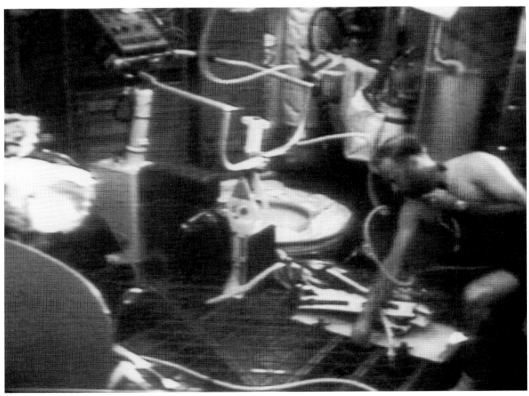

program in space. Dr. Fletcher stated, "It is essential to the future welfare of the country."

After the ceremony, I walked through a test mockup of the Skylab workshop, its inside as roomy as a two-bedroomed house. It had a kitchen, three sleeping cubicles, a living room, a shower, and a fully equipped bathroom. The kitchen was stacked with modern conveniences. Beyond a small table a porthole-type picture window would allow astronauts to look out at the majestic planet Earth as they ate their meals. An important innovation was that Skylab contained nonburnable materials only, the first human habitation to be completely fireproofed: a technology that will be important for bases on the Moon and Mars but, unfortunately, has yet to be applied extensively here on Earth.

There were four flights during the program. The first was to carry the unmanned Orbital Workshop into a nearly circular orbit close to 270 miles (435 km) above the Earth's surface. A two-stage Saturn V launch vehicle was used to do this. Next, flight two carried a crew of three in an Apollo Command/Service Module into an elliptical orbit using a Saturn 1B booster. After separation from the booster the crew used the Service Module propulsion to rendezvous with the Orbital Workshop and dock with it. The crew then transferred to the Orbital Workshop for the orbital mission lasting about one month. On conclusion of the mission, the crew returned to the Command/Service Module for a splashdown landing in the Pacific.

Flights three and four were similar, each mission carrying a crew of three astronauts for operations in the Orbital Workshop for almost two months. The program was highly successful both in gaining information about the medical aspect of microgravity and in experiencing living in a space station of reasonable size; about equivalent to a three-story house. Also, the program produced some extremely important scientific results among which were unprecedented images of the Sun (figure 6.14) and its activity as viewed from beyond the absorption of the terrestrial atmosphere.

The Apollo–Soyuz (figure 6.15) test project was a result of the agreement signed on May 24, 1972 by U.S. President Nixon and Soviet Chairman Kosygin to cooperate in the exploration and use of outer space. An immediate aim was to organize, develop, and schedule a test docking mission in LEO. This first international manned flight took place in 1975, for which NASA developed and built a Docking Module to serve as an airlock and transfer corridor between Apollo and Soyuz. An Apollo Command/Service Module, originally built for the lunar program but no longer needed for that purpose, was modified to include equipment to operate the Docking Module and additional propellants for maneuvering.

The Soviets also modified their Soyuz spacecraft (figure 6.16), particularly in the docking system, so that it could interface with the NASA Docking Module. Soyuz was launched by an A-2, multiengined rocket (figure 6.17). The spacecraft consisted of three modules: an Orbital Module used by the crew for work and rest in orbit, a Descent Module with main controls used by the crew for launch, descent, and landing, and an Instrument Module with subsystems required for power, communications, and propulsion. One benefit to be realized by Apollo–Soyuz was to close the gap in U.S. manned spaceflight between the Skylab missions and the Space Shuttle operations, which were not expected until 1978.

The establishment of the Skylab Orbital Workshop and cooperation with the Soviets with the Apollo–Soyuz linkup in space as important steps toward a human

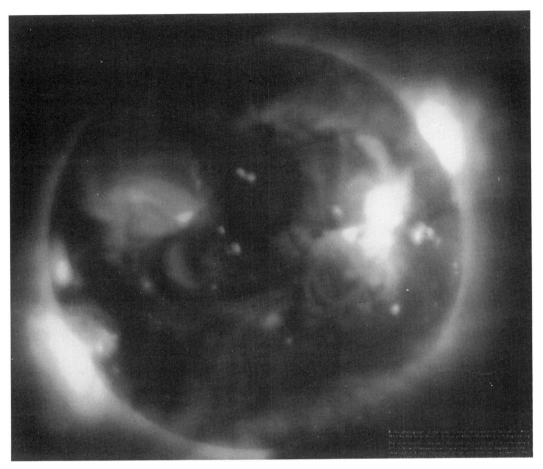

FIGURE 6.14 Skylab obtained some spectacular images of the Sun of a type impossible to obtain from within Earth's atmosphere. This X-ray photograph of the solar corona, the extremely thin outer atmosphere of the Sun, shows loops and arches caused by interaction of the Sun's magnetic field and the ionized gases of the corona. (*NASA*)

future in space was not followed with further development of a sustained U.S. human presence in LEO to keep up with the Soviet presence in this important region. LEO, a region just above Earth's appreciable atmosphere, and thus reachable by spacecraft from Earth's surface with the lowest possible energy cost, will always play a critical role in space operations. It is the natural location for a space station that is to act as a stepping stone to the Solar System, and is expected to contain in the next century an international complex of assembly, research, repair, and production facilities, many in support of operations involving the return of satellites, materials, products, and human crews from high Earth orbits, lunar bases, and interplanetary missions.

Unfortunately all the wonderful plans for a sustained U.S. presence in space fell apart as even the Apollo program ended. Service Modules and Apollo Command Modules became relegated to museums, and the large Saturn space boosters were allowed to deteriorate in storage. Ultimately Skylab fell from the sky in a blazing trail across Australia because there was no way to adjust its orbit and keep it in

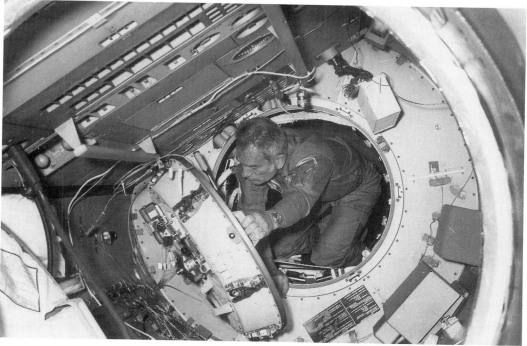

FIGURE 6.15 Apollo-Soyuz was a first U.S. and Soviet joint manned mission in Earth orbit. (a) This rendering by artist Paul Fjeld shows the two spacecraft—Apollo's Command/Service Module, and Soviet Soyuz spacecraft—linked via the Docking Module. (NASA) (b) Astronaut Donald K. Slayton checks out a docking module during preflight preparations at Johnson Space Center. Crews transferred between the two spacecraft through this docking module during the mission. (*NASA*).

FIGURE 6.16 The Soviet Soyuz spacecraft had three main components: a spherical orbital module, the descent vehicle, and the instrument assembly module with its solar panel arrays. (*NASA*)

space. The U.S. effectively retreated from outer space and concentrated on a space shuttle transportation system, which although needed should not have been the sole activity of the piloted spaceflight program. The shuttle was delayed in becoming operational and accordingly was unable to prevent the end of Skylab. In typical penny-pinching activities in connection with the advanced technology of the space program, Congress cut funds from the Skylab program to save the expense of propulsion power to modify its orbit. The rationale, which it turned out was not very rational, was that the Shuttle would be available to protect Skylab's orbit from decaying. Unfortunately the Sun did not cooperate and the effects of its activity on Earth's upper atmosphere increased the drag on Skylab before the Shuttle could be made available to save the first American space station from a fiery death.

In 1982, NASA Administrator James M. Beggs established a space station task force following his earlier statement that a space station should be the major U.S. space goal after the early success of the Space Shuttle transportation system. In two years the task force had defined the requirements for a permanently inhabited space station and a technology steering committee was appointed. A series of international space station user workshops identified many uses to which such a space station could be put, and industry started to conduct analyses for system definition and preliminary design of the station. Many reviews were conducted by various prestigious professional bodies.

In 1984 President Reagan directed NASA to develop a space station within a decade and later he invited other nations to participate. The invitations were warmly received in Canada, Europe, and Japan. Agreements were signed defining what these international partners would contribute in the forms of modules and other hardware. Both houses of Congress were also generally enthusiastic about the program. It provided a new peaceful and worthwhile advanced technology goal for

America and would undoubtedly stimulate the economy just as Apollo had done some 20 years previously. Freedom was confidently expected to stimulate the development of new technologies and advance industrial competitiveness, enhance commercial enterprises in space, and provide facilities to undertake many important scientific research projects impossible to conduct during the short orbital flights of the Space Shuttle. The space station would, moreover, help the U.S. to become preeminent in space at a time when many other nations were becoming increasingly self-reliant in space.

In addition to the practical technological and scientific benefits from establishing Freedom, the symbolic value of a space station is enormous. The station would represent the commitment of the United States to a future in space. It would confirm that the nation did not intend to rest on its laurels like a tired athlete, no longer able to compete in the world scene. With Russia's Mir space station still operating despite all the difficulties that nation faces, Freedom would indicate to the world that the U.S. is not indeed bankrupted by an unprecedented national debt, the legacy of the Reagan years, and unable to lead humanity into a challenging

FIGURE 6.17   This type of launch vehicle was used to send a Soviet Soyuz spacecraft for a rendezvous with the Apollo Command/Service module. (*USSR Academy of Sciences*)

future. It would prove that the U.S. has the vision to seek out and accomplish many operations in space and on other worlds that will be intellectually rigorous, technically extremely demanding, and without doubt economically important to the wellbeing of the world's peoples.

Space station Freedom is a long awaited and overdue program to bring the U.S. back into the long-term human expansion into space. Its technology is essential for this expansion, beginning with the permanent bases on the Moon and on Mars.

• A NEW BREED OF LAUNCH VEHICLES

The amount of material that must be lifted into LEO from Earth for the lunar program, and the size of modules to allow efficient assembly and integration testing of the lunar vehicles at LEO, require space transportation vehicles more powerful than the existing space shuttle. A payload capacity of about 100 tons to LEO is needed. Several alternatives have been suggested for the new transportation system; two of these are a heavy-lift unmanned launch vehicle based on the existing STS shuttle configuration (figure 6.18) and a completely new advanced launch vehicle (figure 6.19). The new vehicle could carry 300,000 pounds of payload into a 500

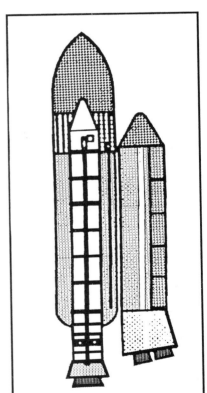

FIGURE 6.18 A return to the Moon via an orbital transportation node will require more powerful launch vehicles than the existing Space Shuttle. One possibility is an upgraded shuttle shown diagrammatically in this sketch. Its advantage would be use of existing shuttle launch facilities.

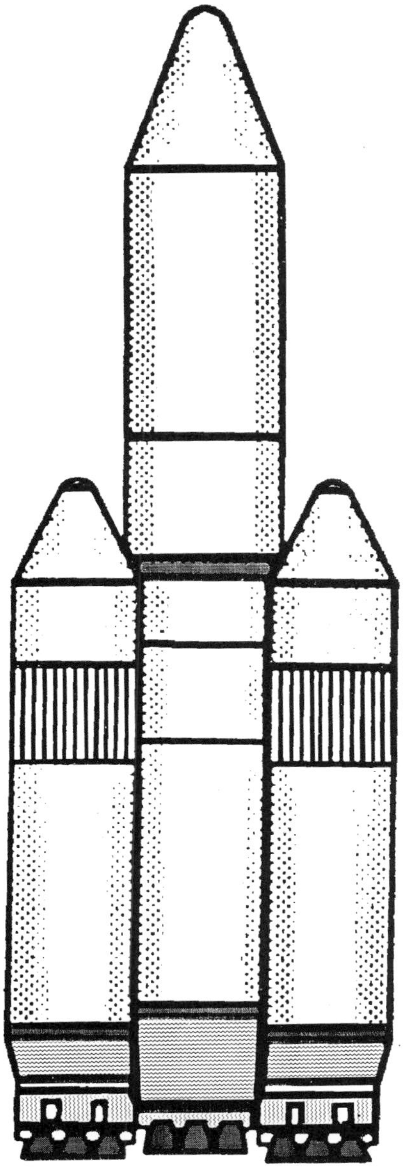

FIGURE 6.19 A preferred option is development of a new advanced launch vehicle of a type similar to that shown in this drawing.

mile polar orbit, or more into an equatorial orbit—more than twice as much payload into LEO as an upgraded and modified shuttle, according to some studies. However, the modified shuttle would have the advantage of being able to use existing launch pads while the advanced vehicle would have the added expense of new launch and recovery facilities. The capacity of the payload compartments of either vehicle will be adequate to carry both lunar transfer and lunar ascent/descent vehicles into LEO without having to assemble them there from components, as would be the case with the existing space shuttle.

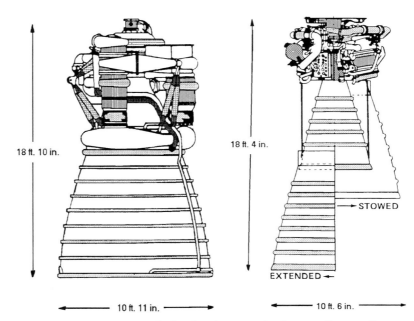

FIGURE 6.20 A major requirement is the development in the near future of a powerful rocket engine pair; a booster engine using liquid oxygen and JP4 fuel, and a main engine using liquid oxygen and liquid hydrogen and having an extendable nozzle to improve vacuum performance. (*After NASA study report*)

The advanced launch vehicle would require the development of a new liquid oxygen/liquid hydrocarbon booster engine and a liquid oxygen/liquid hydrogen upper stage engine derived from today's technology (figure 6.20). Several configurations have been studied as shown in figure 6.21. A 1986 NASA report concluded that a parallel burn, two-stage configuration (figure 6.22) is preferable. The pacing item would be the new high-powered booster engine for the first stage, the development of which should be started quickly. Recovery and reuse of the boosters would cut operating costs, especially if they could be recovered on land rather than in the ocean as is true with the Space Shuttle boosters. The development of the heavy-lift launch vehicle would provide many derivative boost vehicles for a variety of anticipated future space applications (figure 6.23) as well as provide the payload-to-LEO capability required for lunar and Mars missions. As such it would be an excellent investment to safeguard this nation's share of economic benefits to be derived from the inevitable future human expansion into the Solar System.

Another possibility mentioned in recent studies is to revive the Saturn V program; blueprints for the big booster are probably still available. Use of updated technology in many of the subsystems might result in an effective heavy-lift launch vehicle without having to develop a new vehicle from scratch. Also, since the launch vehicle would not be needed for humans since crews would use the space shuttle to travel to and from LEO, the revived Saturn V need not be man-rated so that the cost per flight would be considerably reduced compared with when it was used in the Apollo program. However, use of the Saturn V may not be cost effective if it continued as a throwaway vehicle; that is, if the various stages could not be recovered for refurbishing and reuse. Ideally the heavy-lift launch vehicle should incorporate at least partial and preferably complete recovery and reuse. During the Saturn program there were several studies to see if the components could be recovered, since the big vehicles cost almost $200 million apiece. The relatively small number of Saturn V flights did not warrant the additional expense of developing recovery systems. Should the Saturn V be revived and used to establish and support a lunar base, recovery systems of at least two stages would be necessary to reduce costs.

The heavy-lift launch vehicle program has not yet been funded, but it needs to begin soon either with a new vehicle or a revived Saturn because development to an operational booster could take as long as 15 years with concurrent development of engines, structure, avionics, payload bays, test facilities, launch site, command and control operations systems, and recovery and refurbishing facilities. Without this type of heavy-lift launch vehicle the U.S. will be destined to play only a minor

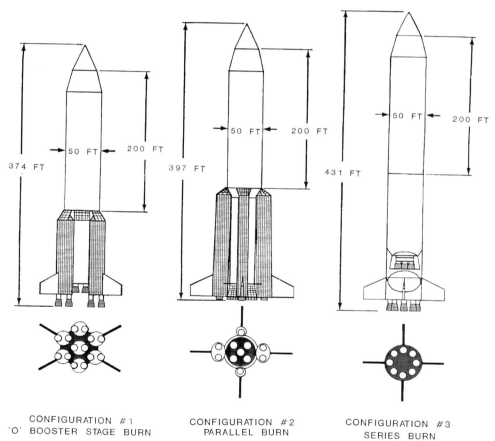

CONFIGURATION #1
'O' BOOSTER STAGE BURN

CONFIGURATION #2
PARALLEL BURN

CONFIGURATION #3
SERIES BURN

FIGURE 6.21 Several options for a heavy lift launch vehicle have been considered. Three configurations evaluated during a NASA study are shown in this diagram.

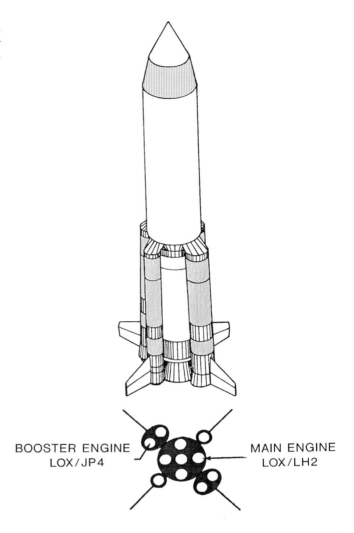

FIGURE 6.22 The configuration preferred was the parallel burn arrangement shown in this diagram adapted from the NASA report.

BOOSTER ENGINE
LOX/JP4

MAIN ENGINE
LOX/LH2

role in future space activities and will have effectively thrown away a quarter century of investment in the space program. The American dream, originated by those Europeans with the vision to use the transportation means of the 16th, 17th, and 18th centuries to establish a new way of life in the New Worlds discovered by the great explorers, will have ended. Other groups of adventurous people will continue the expansion to new ways of life and governance in a widening vision of human derivatives spreading throughout the Solar System and beyond.

• TRANSPORTATION NODES AND LUNAR SPACECRAFT

In one recent NASA scenario for the initial return of humans to the Moon, the transfer and the ascent/descent vehicles can be carried by one flight to LEO and the propellants in two subsequent flights of the advanced heavy-lift launch vehicle.

While checkout and crew transfers and propellant loading will be carried out at space station Freedom initially, a full-fledged lunar program will require another space station to act as the transportation node since the activities of the lunar program would seriously interfere with the scientific and other objectives of Freedom's continued long-term operations.

There are several ways in which the mission to establish a permanent outpost on the Moon can be implemented through use of a transportation node in low Earth orbit. The node acts as a connecting link between spacecraft used to transport crew and materials from the Earth's surface to LEO and spacecraft used to transport payloads of people and materials to and from the Moon.

The transportation node will most probably be separate from the space station Freedom, but may be positioned to move around Earth in the same orbit as the space station.

Basically the transportation node has different functions from the orbiting space station. It must be capable of handling and storing large quantities of rocket propellants. It must provide for frequent interactions with a large number of space vehicles of different designs to provide them with docking and other services such as maintenance and propellant loading, as well as transfer facilities for personnel. These activities would not be acceptable to the environment of space station Freedom and would interfere with its scientific missions.

NASA's 90-day study on human exploration of the Moon and Mars suggested a transportation node based on Freedom for the initial phases of the lunar base

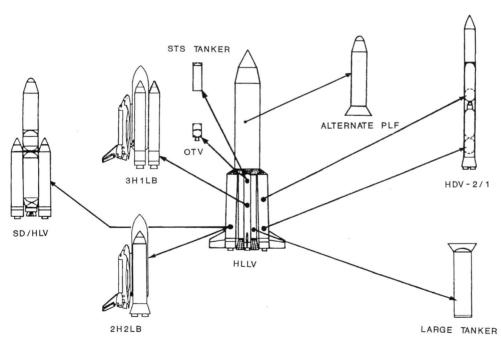

FIGURE 6.23 The development of the various components for a heavy lift launch vehicle of this type would provide several opportunities for these components to be used to upgrade other space systems as shown in this diagram adapted from the NASA study.

operations. This would provide a node to verify the flight configuration of the Lunar Transfer Vehicle and for the expandable and reusable Lunar Transfer Vehicle operations later. Looking further into the future, scientists and engineers at NASA's Johnson Space Center studied several more advanced configurations for the transportation node. Three of these were the Platform, the Drive Through, and the Atrium (shown diagrammatically in figure 6.24) together with the node based on Freedom.

The augmented Freedom configuration builds on the currently approved space station to provide the lowest cost transportation node if not the most effective in the long term. The platform configuration emphasizes processing of the spacecraft within the transportation node to minimize the size of the node structure. The

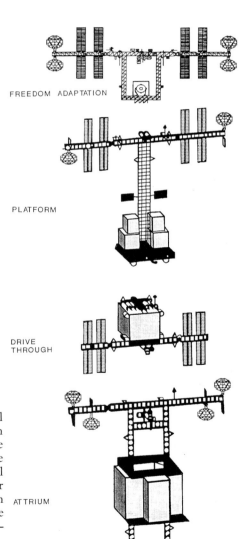

FREEDOM ADAPTATION

PLATFORM

DRIVE THROUGH

ATRIUM

FIGURE 6.24 Human expansion into space will rely greatly upon establishment of a transportation and refueling node in low Earth orbit (LEO). There are several ways to do this. One would be to use the proposed space station Freedom with additional components as shown in the top sketch. Three other versions have been suggested in a NASA-Johnson study. These would be separate from Freedom. The platform would be developed from the Freedom design. Below that is a Drive-Through configuration, and the bottom configuration is the Atrium design. (*After NASA diagrams*)

Drive Through and the Atrium configurations emphasize optimizing the operation and maintenance of the transportation node and its control and stability.

There are two major aspects of operating a transportation node for the lunar outpost program: internal and external functions. Internally these functions include the movement of payload, assembly of the spacecraft, and loading the spacecraft with propellants to prepare it for its journey to the Moon. Externally the functions involve maintaining the correct orientation and position of the transportation node and ensuring that it remains in the appropriate orbit for the extended operations needed to establish and maintain a lunar outpost.

Space station Freedom, in one NASA study, would evolve through four configurations, which would require that capabilities of expansion for the future missions, particularly for additional power generation, must be incorporated in the initial design. The Russian MIR station has been developed through part of such a sequence with more developments planned for the future. Added to Freedom must be capabilities to store and load propellants, assemble and service aerobrakes, process various vehicles, and check out those vehicles. A service track assembly added to Freedom would be needed to provide a fixture for assembling, mounting, and servicing the lunar vehicles. The truss structure of Freedom would have to be augmented with keels and a lower boom as shown in figure 6.24.

Derived from the basic design of space station Freedom, the Platform configuration is shown in figure 6.25. A single long boom contains the solar cell arrays at either end with habitation and workshop modules at the center. A vertical keel ends at a platform on which the lunar spacecraft are assembled. Hangers that provide 880,000 cubic feet (25,000 cubic meters) of volume are at either side of the keel mounted on the platform together with propellant tanks to hold 364 tons of cryogenic liquids (sufficient for two lunar missions). The configuration has a mass of about 300 tons. The control system is passive; it uses the stabilizing torque of the Earth's gravity.

The spacecraft are assembled on the platform and moved by a mobile servicing center traveling along the vertical keel structure. There is an upper docking ring near to the habitation module to which the lunar vehicle is moved. There the crew boards the spacecraft and it departs on its journey to the Moon. Two complete lunar stacks can be assembled in this version of the transportation node.

The Drive-Through configuration is illustrated diagrammatically in figure 6.26. It has an extended boom with solar cell arrays at either end. At the center section of the boom a hanger structure is mounted, below which are crew quarters and a space shuttle docking area. Propellant storage tanks are attached outside the cubical hanger. Stretching from the crew quarters is a tunnel leading to a control module inside the hanger space. The unpressurized hanger measures 165 feet (50 m) by 115 feet (35 m) by 82 feet (25 m). It can house two mission spacecraft at once. Each spacecraft, sometimes referred to as a stack, consists of an orbital transfer vehicle, two aerobrakes, a lunar ascent/descent vehicle, and a lunar payload. The prefabricated modules for the spacecraft are brought initially from Earth into the hanger through the back door. The stacks are attached to fixtures that allow 360 degrees of rotation for assembly and servicing. These fixtures have an access tube that can be connected to the lunar crew module or payload compartment to allow access of crew and maintenance personnel from the transportation node's crew quarters in a

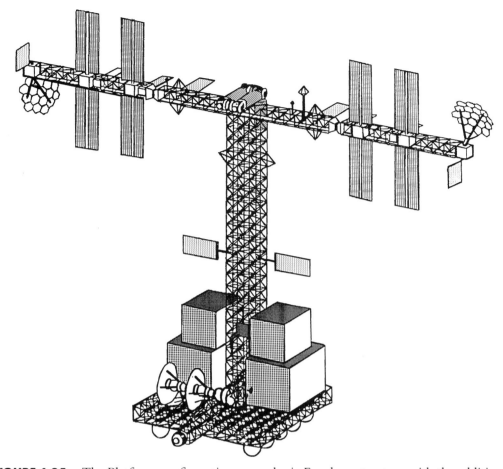

FIGURE 6.25 The Platform configuration uses a basic Freedom structure with the addition of a vertical keel with a platform at its Earth end on which the lunar spaceships are assembled or serviced, as described in the text. (*after NASA-Johnson study*)

pressurized environment, thus avoiding use of spacesuits and extravehicular activity (space walks) for much of the refurbishing activities.

When a stack is ready for a mission to the Moon it is moved to the forward fixture where it is loaded with propellants before leaving the hanger through the front door. A second stack is readied on the back fixture to be available for rescue missions, if needed, or in preparation for the next scheduled lunar mission.

The propellant storage facility would consist of four tanks capable of storing 182 tons of liquid oxygen and liquid hydrogen cryogenic propellants. Additional propellants also can be stored in heavy-lift tankers docked with the transportation node if required.

The crew quarters, or habitats, can house a permanent transportation node crew of six, as well as three crew members from the Earth-to-LEO shuttle, and four lunar spacecraft crew members. There are also workshop modules attached to the crew quarters, and a module from which the logistics of the node are controlled.

A typical Atrium configuration developed in the NASA-Johnson study for a transportation node is illustrated in figure 6.27. Six storage hangars surround an assembly area strung beneath the long boom of the transportation node. The center of the boom carries the crew habitat, workshop modules, and the logistics operations module for the node. On either side of the assembly area is a hangar suitable for housing an orbital transfer vehicle with its aerobrake. The assembly area is large enough to accommodate various lunar vehicles. Two hangars are used to store and maintain lunar ascent/descent vehicles. A fifth hangar holds the lunar modules for cargo and crew, while the sixth hangar services orbital maneuvering vehicles. The hangars are enclosed, as on the other NASA-Johnson concepts of transportation

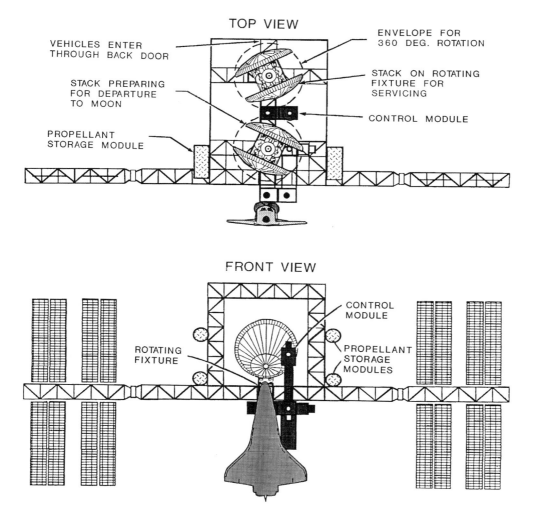

FIGURE 6.26 The Drive-Through configuration (see description in text) assembles lunar spacecraft or services them as they enter from the rear and move through to exit at the front ready for the voyage to the Moon. (*after NASA-Johnson study*)

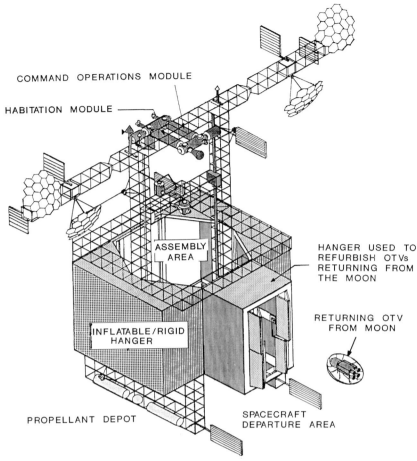

COMMAND OPERATIONS MODULE

HABITATION MODULE

ASSEMBLY AREA

HANGER USED TO REFURBISH OTVs RETURNING FROM THE MOON

INFLATABLE/RIGID HANGER

RETURNING OTV FROM MOON

PROPELLANT DEPOT

SPACECRAFT DEPARTURE AREA

FIGURE 6.27 The Atrium configuration has inflated rigid hangars in which spacecraft are assembled or serviced as described in the text. (*after NASA-Johnson study*)

nodes, to protect the spacecraft and personnel from micrometeoroids and orbital debris.

In the Atrium configuration the six hangars enclose an unpressurized volume of 3.1 million cubic feet (88,000 cubic meters). The mass of the transportation node is 320 tons, and 182 tons of cryogenic propellants can be stored within it.

Extending across from the dual keel structure, and below the main boom that supports the solar cell panels, is the mobile transportation boom. It is used by the crew to move personnel and payloads between the habitation modules and the assembly and departure areas. Spacecraft departure is from the bottom of the assembly area. In one design the bottom of the atrium is open to allow light reflected from the Earth to be used for illuminating the interior. The habitation module provides quarters for six permanent and seven temporary crew members.

A lunar transfer vehicle (LTV) transports personnel and freight between the transportation node in LEO and low lunar orbit (LLO). A lunar excursion vehicle (LEV), sometimes known as a lunar descent/ascent vehicle (LDAV), transports similar payloads between LLO and the lunar surface. Both vehicles can be economically optimized by using a common design to transport either crews or cargoes.

In one configuration suggested by NASA-Johnson researchers in 1989, the LTV is envisioned as a one-and-a-half stage vehicle made up of a reusable core component and expendable propellant tanks (figure 6.28). Some studies have shown that by dropping off and abandoning the propellant tanks when they are exhausted of propellants, originally proposed by Kenneth W. Gatland at the second International Astronautical Congress in 1951 for a minimum satellite vehicle, about 10 percent of propellant load can be saved when compared with a single stage LTV in which the complete vehicle including its tanks is reusable. The core has a dry mass of about 8 tons, and it carries a propellant load of 7 tons and a crew capsule weighing just over 8 tons. The jettisonable propellant tanks have a dry mass of about 6 tons and carry 130 tons of propellant.

The crew module (figures 6.29 and 6.30) can support four crew members for four days on the outward trip to LLO and a further seven days for the journey back to LEO. It contains a galley, a zero-gravity toilet, and provisions for personal hygiene. Shuttle-type hatches permit transfer into the module in a shirtsleeves environment. Suits are also provided for extravehicular activity if needed. To do this, however, the crew module has to be depressurized because no airlocks are provided. Sufficient gas is stored to permit two extravehicular activities in emergencies. The crew is shielded from solar flare activity by water, which is jettisoned before return to LEO to reduce the load on the aerobrake system.

Return to LEO makes use of aerobraking, which significantly reduces the amount of propellant that has to be transported from Earth's surface to LEO for the mission

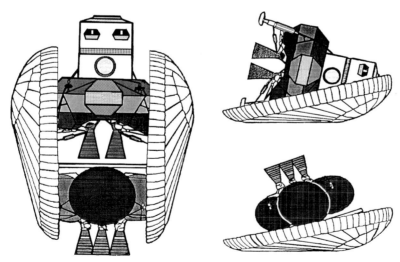

FIGURE 6.28   A possible combination of spacecraft for a lunar mission, described as a stack, is shown in these sketches derived from a NASA-Johnson study. At bottom is a propulsion module consisting of propellant tanks and rocket engines. Above it is a landing module consisting of a propulsion unit with landing legs topped by a crew module. On either side are the aerobraking modules used when returning from the Moon to LEO.

FIGURE 6.29 This diagram adapted from a NASA-Johnson report shows another possible set of lunar spacecraft considered in a NASA-Johnson study. Top is an orbital transfer vehicle with a cargo lander outbound from LEO to the Moon. Beneath it a crew module is substituted for the cargo on the outbound trip. The third configuration is that of an orbital transfer vehicle and crew module returning from the Moon to LEO, while the bottom configuration shows an orbital transfer vehicle and propellant tanker being returned to LEO.

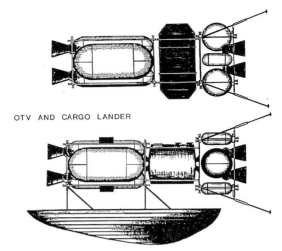

OTV AND CARGO LANDER

OTV AND CREW LANDER

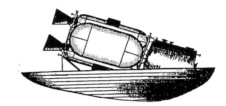

OTV AND CREW RETURNING TO LEO

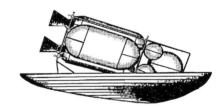

OTV AND PROPELLANT TANKER RETURNING TO LEO

(see figure 6.29). The rigid structure of the aerobrake consists of composite materials protected by advanced thermal coatings. The aerobrake can be used for several returns before having to be refurbished or replaced.

The lunar excursion vehicle (LEV) consists of a propulsion system, landing legs, a crew cab, cargo pallets, and control and communication systems (see figure 6.30). It carries a crew module of about 4.5 tons, a propellant load of 22.5 tons, and has a dry weight of just under 6 tons. It is a multipurpose vehicle, for it not only provides propulsion down to and up from the lunar surface but also can be based unattended on the surface or left unattended in LLO for periods of inactivity. It can carry a crew of four and a cargo of up to 15 tons to the lunar surface when piloted; if it is in an expandable cargo only mode, it can transport 33 tons of cargo to the lunar surface.

The crew modules for LTV and LEV use a common design. However, the LTV has a radiation shielded section to be used if solar storms occur during the transfer from LEO to LLO. The LEV does not have such shielding. The LEV also is designed to be operated under zero gravity and lunar gravity conditions. The LEV has two console positions for crew members to operate the landing and touchdown sequence in tandem, much like the pilot and co-pilot of an aircraft.

Several concepts have been identified during recent studies for habitats to be used on the lunar surface, first as an outpost and later as a permanent base. Power systems, vehicles, and life-support systems to support the habitats have also been reviewed and some designs have been analyzed in detail. Habitats are of two basic types: those that are modular and can be transported to the lunar surface with proposed LTV and LEV, to be assembled into an integrated group by placement of the prefabricated units on the surface, and those that are constructed on the surface from various units brought from Earth and by use of lunar materials.

All the equipment used on the Moon—the habitats, life-support and environmental control systems, power supplies, vehicles, construction shacks, laboratories, and logistic and maintenance modules—will act as important test configurations for later use on Mars and elsewhere in the Solar System later in the next century, such as on satellites and asteroids.

Initially the first outpost on the Moon as well as the transportation node probably will be derived from space station Freedom technology; hence the importance of the space station program. The technology developed for Freedom will apply for the outer structure of a pressurized module, or modules, that will be set on the lunar surface. This probably will be a simple cylindrical module about 30 feet (8.2 m) long and nearly 15 feet (4.45 m) in diameter (figure 6.31). Attached to it will be an almost identical module that will be used as a laboratory, also derived from Freedom technology. Plastic bags will be filled with lunar regolith and draped over

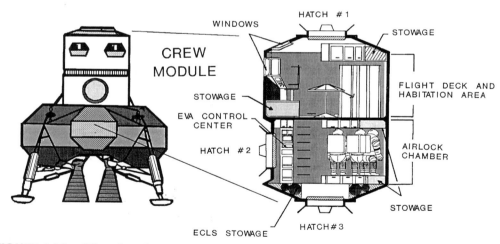

FIGURE 6.30 The piloted vehicle shown in this adaptation of a NASA diagram consists of a landing vehicle and a crew module. A cargo module could be used instead of the crew module in an unpiloted version. The crew module has two levels; a flight deck and habitation chamber, and an airlock and storage chamber. Three hatches allow crew members to exit the crew module in various ways; direct descent to the surface or exit to a vehicle from the airlock chamber, or direct exit from the top of the flight deck if required.

FIGURE 6.31 Initial modules of the lunar outpost will probably be prefabricted cylinders such as this construction shack. The shell is cut away to show the interior working quarters. The module will be covered with bags of regolith to protect it from the lunar environment. (*NASA-Johnson*)

the cylindrical modules to protect them against radiation, micrometeoroids, and thermal extremes during the long lunar days and nights.

If Freedom continues as a viable program despite much unwarranted criticism which parallels very closely the criticisms by the scientific and political community of the Apollo manned program during the 1960s, the regenerative life-support system developed for Freedom will be used in the habitat and laboratory modules on the lunar surface. As planned for Freedom, this life-support system will recover 90 percent of oxygen from carbon dioxide and will reclaim potable water from hygiene and waste water. The Freedom system also will control temperature and humidity, atmospheric composition and pressure, and will provide storage for food and facilities for trash handling, dish washing, laundry, and personal hygiene.

• CONSTRUCTING THE FIRST OUTPOST

The establishment of a lunar outpost will demand a full understanding of closed ecological systems that will require experiments with such systems here on Earth. Already some steps have been made in this direction with experience gained from nuclear submarines, bases in Antarctica, specialized research undertaken by aerospace companies for the Skylab program, and some private enterprise undertakings to check how humans can survive in totally enclosed environmental systems. The lunar outpost program will require also the development of advanced teleoperated and automated rovers capable of performing prospecting and unmanned exploratory roles on the lunar surface much as was done by some Soviet lunar rovers during the Apollo era. Rovers of this type will have applications when we explore other planetary surfaces such as that of Mars.

Preliminary studies are needed for the selection of sites for the lunar outpost. One study by NASA Johnson Space Center, published in 1989, suggested a site at Lacus Veris, a location on the western limb region of the Moon on smooth lava plains within the concentric mountain ranges surrounding Mare Orientale (figure 6.32). Other sites that had been suggested are in the Mare Nubium, the Taurus-Littrow valley that was explored by Apollo 17, and the South Pole. The latter is not really suitable for the initial outpost because it would lead to more difficult navigation and increased energy requirements.

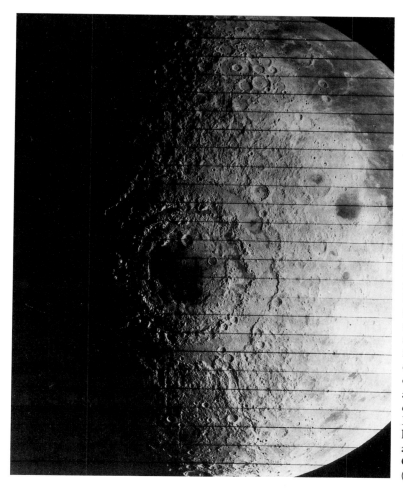

FIGURE 6.32 A possible site for the establishment of the first lunar outpost has been suggested as one in the Lacus Veris region of Mare Orientale, a region just inside the third concentric ring of mountains. On this photograph it is the small dark region just below the crater with the central mountain at the two o'clock position on the third mountain ring. Peaks in this mountain ring, the Cordillera Mountains, rise 20,000 feet (6,100 m) above the surrounding terrain, and are the highest lunar mountains. The site is close to the limb region of the Moon and gives access to the far side and direct line-of-sight to Earth for parts of each lunation. The dark area at upper right is part of Oceanus Procellarum. (*NASA*)

Lacus Veris is at latitude 13 degrees south and longitude 87.5 degrees west. The site is close to many features of great geological interest; it provides a smooth landing area of sufficient extent to separate landing and launch pads from the lunar habitat and science laboratories so that the latter will not be affected by the dust raised by spacecraft landing or taking off. From the site, also, exploring crews can travel to the far side of the Moon and should be able to set up initial astronomical observatories screened from light and radio pollution of Earth. Even if the initial outpost does not have the resources to set up small observatories away from the outpost, the outpost itself, because of the Moon's libration, will be screened from terrestrial radiation pollution for eight days during each lunation period. Except for those eight days, the site allows line-of-sight visibility and communication with Earth. Based on Apollo orbital geochemical observations, the Lacus Veris area should provide oxygen-yielding minerals, such as ilmenite.

Design of the outpost itself is based on locating the landing and launch pads about 1.5 miles (2.5 km) away from the habitats and laboratories. The NASA study suggests placing several landing areas located on a north-south line since spacecraft will come in around the near side of the Moon in a flight path traveling from the east to their landing. Takeoff back to lunar orbit will be toward the west and around the far side of the Moon. The landing sites to begin with will be on an unimproved flat area selected to be relatively free of obstacles. As the outpost is developed, the landing sites will be improved and paved with the gravel resulting as byproducts from the processing of lunar rocks to extract lunar resources from them. Suitably placed navigation beacons will be used to navigate the spacecraft to the landing site so that even cargo-carrying vehicles can land safely. As the movement of personnel and freight becomes commonplace, services and refueling capabilities will be established at the landing site, which will eventually become a spaceport similar to an airport, with several types of vehicles to transfer freight and cargoes from the site to the main outpost habitable area.

At the habitat, which will ultimately be accessed from the landing site by a gravel road, there will be several specialized modules; a habitat, construction shack, laboratories, maintenance facilities, power generating plant, oxygen extraction plant, command, control, and logistics modules, and vehicle maintenance facility (figure 6.33).

Transportation vehicles will be needed for operation between landing site and habitat; these vehicles will be designed to unload freight, including complete modules, from cargo-carrying landers, and crews from manned spacecraft. These operations will require several types of vehicle of modular design (refer to figures 6.2 and 6.9). To a wheeled and powered chassis, various modules would be attached: flat bed for complete logistics modules brought by space freighters from LEO, pressurized and open trucks to move various types of cargo, some of which may require refrigeration or other protection from the lunar environment, and busses to move people and light cargoes. As the lunar outpost becomes larger, transportation vehicles will be needed that will allow crew members to avoid having to don spacesuits to exit from their spacecraft into the surface vehicle. The pressurized interior of the transportation bus will allow crew members to travel in a shirtsleeves environment to the habitat without being exposed to the near vacuum environment of the lunar surface.

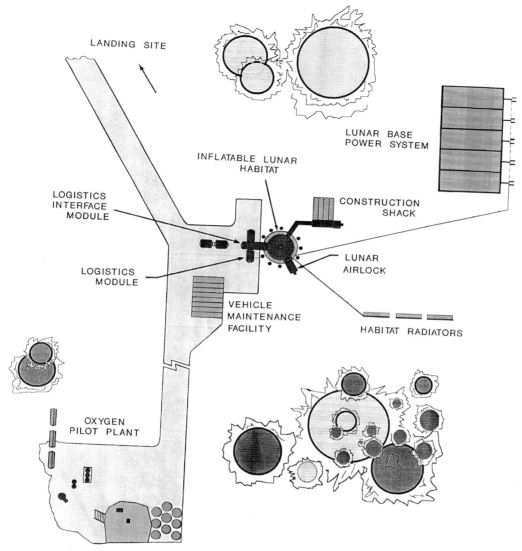

FIGURE 6.33 Plan of a lunar outpost as suggested in a NASA-Johnson study report. The various units described in the text are identified on the drawing.

At the beginning of establishing the lunar outpost, the first crews will operate from the landing craft, as they did from the Lunar Module of Apollo. In the NASA-Johnson scenario the first simple module of the lunar outpost will be a construction shack. From this base module crews will go out on the lunar surface to install or assemble the other units of the outpost. The construction shack (figure 6.34) is a self-contained unit. Setting it up at the site for the main habitat reduces travel to and from the landing site to the habitat site during construction of the habitat and installation of its supporting units.

The construction shack would provide living quarters, solar power generators, an environment control and recycling system, and a small workshop (figure 6.35).

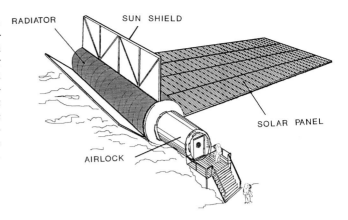

FIGURE 6.34 This drawing, adapted from the NASA lunar outpost report, shows a construction shack with an airlock and a power generating solar panel. The sun shields are moved to protect a radiator from direct sunlight or heat radiated by the lunar surface. The radiator is needed to get rid of excess heat generated in the module by humans and equipment operating within it.

Its cylindrical unit some 50 feet (15 m) long and 12 feet (3.7 m) in diameter, could be partially submerged in the regolith, stabilized by outrigger footpads, and covered with bags of regolith for thermal and radiation protection. Airlocks at both ends would provide access to the lunar surface (figure 6.36).

Construction of the permanent outpost will occupy many hours of work on the surface, requiring longer periods of duty on the surface than could be provided by crews operating from the lander spacecraft. Since the construction shack would be delivered to the lunar surface as a completed module, fully equipped with living quarters and life-support and other necessary systems, it would require minimum effort to activate on arrival of the first crews. Mobile equipment would remove it from the landing vehicle and transport it to the habitat site. This equipment initially might be multipurpose, similar to some terrestrial construction machines. It would be equipped, for example, with excavation tools to prepare the site for the construction shack and to load the protective bags of regolith. Again, design might be modular with a small four-wheeled unit that can easily be modified for different tasks. For example, a simple U-shaped cradle on several wheeled units operating as a train could provide support for the construction shack cylinder when transporting it from the lander to the habitat site. These cradles could then be removed and other equipment affixed to each wheeled vehicle, such as a crane with outrigger footpads (figure 6.37). The lunar gravity of one-sixth terrestrial gravity will make many lifting and moving tasks much easier than they would be on Earth, although it must be remembered that while the weight of an object may be considerably reduced, its momentum will remain the same. While much less energy will be required to lift or lower a given mass, the same amount of force as on Earth will be required to start it into motion or stop it from moving.

When the outpost becomes fully operational, the construction shack could be modified for other purposes; for example, to house workshops or laboratories. Also, similar modules, based on the same design, could be used to expand human activities to other locations on the Moon at which a fully operational outpost will not be needed.

With the construction shack established and activated, the main task of erecting the lunar habitat to provide a living and working environment for the outpost's crew of 12 men and women can begin. A NASA study suggests one approach would be to use an inflatable construction. It would be based on a packaged spherical envelope that would be inflated at the site of the outpost. The spherical habitat would be about 50 feet (16 m) in diameter, sufficient to enclose four floors of operational space and suitable to support the crew of 12 permanent personnel. Temporary visitors on specialist assignments could be housed in the construction shack living area. A central vertical shaft in the habitat sphere will provide access to the various floor levels. These modular floors would be assembled from prefabricated units and supported on telescopic columns with a headspace of about 10 feet (3.2 m) on the floors with larger areas, and a somewhat greater height at the top and bottom levels where the curvature of the enclosure restricts useful square

FIGURE 6.35 Crew members are shown monitoring activities at the lunar outpost in this cutaway of a construction shack. One is watching a television view of activities, the other is exercising. To the right of that crew member are booths for controlling teleoperated vehicles and equipment. (*NASA-Johnson*)

FIGURE 6.36 The drawing shows the location of one of the airlocks though which crew members gain access to the lunar surface from cylindrical modules at the lunar outpost. (*NASA-Johnson*)

FIGURE 6.37 Small four-wheeled vehicles could be modified to accept several types of construction equipment. Here a crew member is using a crane, stabilized by outriggers, to gather lunar regolith which will be used for various purposes such as protecting and insulating the habitats. (*NASA-Johnson*)

footage. Mission operations would be on the first level, below which would be a storage area. The second level would be base operations and above that the crew support service level. Top level would be allocated to quarters for the crew.

The spherical envelope would withstand an internal pressure of normal terrestrial sea-level pressure (14.7 psia). Its outer surface would be coated with a thermal protection material. This pressurized shell would bear the loads imposed by the difference between internal pressure and the near vacuum outside. The structural loads of the floors and the equipment on them would be borne by the columns of the cylindrical core and the radial beams of the floor structure on which the modular floor panels are mounted.

To install the habitat, crews operating from the construction shack will excavate a suitably shaped depression in the regolith, possibly modifying an existing small crater at the chosen site. Piles will be driven into the bottom of the excavation to be used as anchors for the habitat. The packaged habitat will then be anchored in place in the excavation and inflated to its spherical shape. Any voids left between the sphere and the excavation will be filled with regolith. A pressurized tunnel is connected between a port on the inflated sphere and the construction shack. The whole is shielded with regolith, and installation of the interior structure and equipment can then proceed in a shirtsleeves environment; the workers would operate out of the construction shack through the pressurized tunnel.

An airlock will be connected later between the habitat and the lunar surface to allow crew members to don or discard spacesuits in moving from the interior of the habitat onto the surface, or vice versa. The airlock will contain dust removing facilities for returning crew members before they enter the pressure changing crew chamber.

To support space station Freedom, NASA has conducted research on the design of new spacesuits that are highly reliable, provide greater flexibility to the wearer, do not require much maintenance, keep a constant internal pressure despite how the suit is flexed, and shield the wearer against radiation and the impact of small meteoroids. Two versions of the suit have already successfully passed many tests. While the Freedom suit is what is known as a hard suit, much of this new technology can be applied with modifications to suits designed for support of the lunar outpost, its installation, and subsequent operation. One innovation, for example, is a suitport that gives the crew member the ability to put on or take off the suit in a few seconds compared with several minutes for existing spacesuits. Also the suit need not be brought into the living environment but can be entered through a hatch (figure 6.38) with the suit always remaining outside the living environment. Thus lunar dust would not be carried into the habitat or into vehicles. The suit also can be pressurized to Earth's surface pressure so that its wearer will not need to breathe pure oxygen for several hours prior to wearing it in order to avoid the "bends."

Also attached to the habitat will be a group of modules for logistics. One of these, for example, serves as a berthing port and loading dock for pressurized rovers and transporters. Another module might provide an additional airlock for access to the lunar surface, an important backup should the main airlock from the

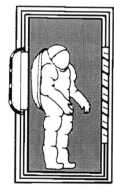

FIGURE 6.38 A suitport developed for space station Freedom will be useful for crews operating at the transportation nodes in LEO. It might be adapted also for use at the lunar base. The suitport allows an astronaut to enter or exit from a spacesuit quickly. On the Moon it might be very important because it would avoid carrying lunar dust into areas where it might interfere with machinery or experiments or prove troublesome to humans.

habitation develop problems. An interface module may be used to connect these modules and provide a channel to the habitat. Figure 6.39 shows a plan and side elevation of the habitat and its accompanying units as they might be emplaced on the lunar surface. The maintenance hangar (figure 6.40) is open to the lunar environment but offers protection from solar heating. Within this hangar crews can assemble and service vehicles and equipment for use around the base and on exploratory missions.

To begin with the lunar outpost will operate probably on about 100 kilowatts of electricity, but as the various modules are installed and come into operation, the

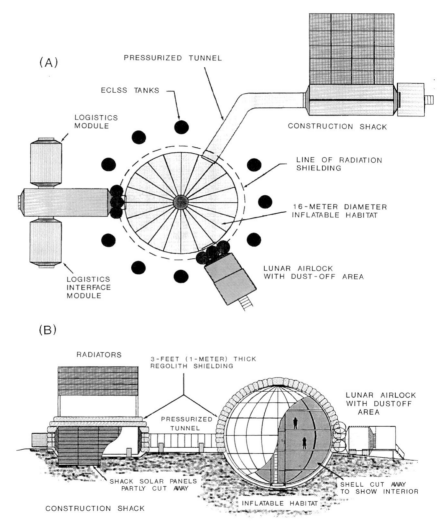

FIGURE 6.39 At NASA Johnson Space Center a team developed a design concept for the habitation module of the lunar outpost. (a) Plan of the module described in the text shows the supporting construction shack, logistics modules, and airlocks. (b) An elevation illustrates how the habitat is placed within a depression on the lunar surface, is covered with protective bags of regolith, and has various floor levels within its inflatable structure. The different levels are accessed through a central vertical shaft.

FIGURE 6.40 A hangar open to the lunar surface will be used for assembling and maintaining equipment and vehicles as shown in this artist's impression. (*NASA-Johnson*)

power demands will increase rapidly. Initial power demands to support the environmental control and life-support system, tools for construction, recharging of vehicle and spacesuit batteries, and communications can be satisfied with conversion of solar radiation either by photovoltaic or photodynamic methods (figure 6.41). The former uses an array of semi-conductor solar cells to convert sunlight directly into electricity. Unfortunately, it is not very efficient, converting only between 5 and 10 percent of incoming solar energy. Although it lacks moving parts, the system must have a storage capacity of batteries or fuel cells to supply electrical power during the long lunar night. To supply 100 kW a solar array must have a relatively large area, about 21,500 square feet (2,000 square meters). Such a large area may have to be obtained by shipping the solar arrays to the Moon as "carpet rolls" and unrolling them on the lunar surface (figure 6.42). The energy per unit mass is low, requiring an array with a mass of about four tons to deliver 100 kilowatts of electricity. A great advantage of this type of solar converter is that it uses off-the-shelf technology.

The solar dynamic system also requires means to store energy for the 14-day lunar night, but it is a more efficient converter of solar radiation energy into electrical energy, so it requires a smaller collector area. However, it requires a much greater mass—ten times the mass of the photovoltaic system—for an equivalent yield of electrical energy. This system, which was incidentally proposed for use in

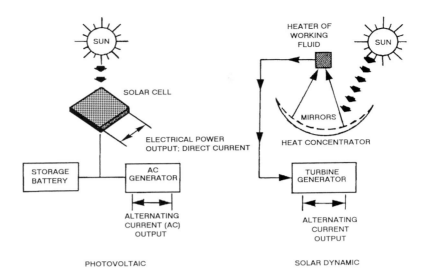

FIGURE 6.41 Two methods of converting solar radiation to electricity in space or on the Moon are shown in this diagram. A photovoltaic system uses solar cells and a DC to AC converter. A closed cycle dynamic system concentrates solar radiation to heat a working fluid which is used to drive a turbine connected to an AC generator. Both methods would need some means to store energy for use during the long lunar night. The photovoltaic system might use a battery or a regenerative fuel cell. The dynamic system might store the energy in a sufficiently large and heat insulated reservoir of the working fluid.

the Ross space station of 1948 and in the author's proposed unmanned communication satellite system of 1949, and the von Braun wheel-shaped station of 1952, uses a thermodynamic cycle to convert the heat from sunlight into useful work. The sunlight is concentrated by a mirror surface to heat a working fluid to a high temperature. This working fluid expands through a turbine that drives an electric generator. Because the system has moving parts and flowing, heat-exchanging fluids, it is more complex than the passive solar cell system and will require more maintenance or servicing.

Both systems must be protected from lunar dust that would otherwise reduce the amount of incident sunlight reaching the conversion system. Both systems will require energy storage equipment, probably regenerative fuel cells, to supply electrical power at night. The fuel cells will store energy during daytime by using surplus electricity to electrolyze water into oxygen and hydrogen. At night the oxygen and hydrogen are recombined to form water thereby releasing energy that is converted into electricity. Fuel cells have been used in the Apollo program and have the advantage over batteries of being only one-tenth the mass of batteries for an equivalent storage capacity. However, the dynamic system has an advantage over the photovoltaic system in that the energy extracted from sunlight is in the form of heat, and heat can be stored more easily than electricity. The dynamic system could heat reservoirs of working fluid that could be stored in underground insulated tanks to be used with the turbines during the lunar night. This would considerably reduce any requirement for fuel cells to store energy. Since both systems are to be investigated for use in space station Freedom, the most appropriate technology will be available for the establishment of the lunar outpost from experience gained with the space station.

As the outpost expands to a full operational capability solar sources of electricity will be insufficient to meet the energy needs. Nuclear sources will be necessary. Radiation shielding can be obtained by burying the reactor below the lunar surface (figure 6.43). Extremely large amounts of electrical energy, measured in megawatts rather than kilowatts, can be generated continuously so that batteries or fuel cells are not needed, except possibly for backup in case of downtime of the nuclear system to service it. While it will not be feasible to service the reactor core itself, the supporting controls and waste heat radiation system may need periodic maintenance. Ultimately, as the core material deteriorates because of build up of fission products, the reactor core will have to be abandoned and another installation set up using much of the control and heat radiation system from the old reactor. But many decades of service will be obtained before such a change will be needed. A nuclear system also has the advantage of producing more power per unit mass than

FIGURE 6.42 Artist's impression of a field of solar cells that are unrolled carpet fashion on the lunar surface to provide electrical power for the initial outpost. In the foreground, to the left, are two fence-like structures forming part of a number of such structures which are used to radiate excess heat generated by the outpost modules. (*NASA-Johnson*)

FIGURE 6.43   Large lunar bases will require establishment of a nuclear power station on the Moon with the nuclear reactor buried beneath the surface and a radiating pattern of thermal radiators, as shown in this artist's rendering for NASA. (*NASA-Johnson*)

the solar systems. This is important in view of the large amount of initial mass of propellant required to transport each pound of freight from Earth to the Moon.

An important aspect of all operations on the surface of the Moon is that facilities and equipment must be provided with thermal control systems, to prevent the loss of heat during the long lunar night and to dissipate excess heat during the searing lunar day. Because the Moon is airless, excess heat or waste heat must be radiated; convection and conduction are inappropriate on the lunar surface. This requires installation of large areas of radiators that must compete with the heat being radiated from the hot sunlit lunar surface, since radiators will absorb this heat from their surroundings. While increasing the area will dissipate more heat, the large area has to be reduced at night to avoid loss of too much heat. In daytime radiators may have to be shielded from sunlight and from radiation from nearby "hot" lunar surfaces. The sunshields must be automatically controlled to change position as the direction of the sunlight changes during each day.

Lunar dust from the regolith acts as an insulator. If it is allowed to coat radiators it will reduce their efficiency. Means must be provided to keep the radiating surfaces free of this lunar dust that unfortunately has the unwanted and annoying characteristic of sticking to all surfaces exposed to the lunar environment. One proposal from NASA Johnson Space Center envisages a large radiation field with parallel banks of radiators looking very much like a series of fences arranged over the landscape of the Moon. Heat is carried to the radiators from the habitat and other outpost modules by a liquid similar to a closed-cycle air conditioning system. This also presents a problem on the Moon because of the rain of micrometeorites that might puncture the radiators and allow the fluid to leak into the surrounding vacuum. So means have to be provided to detect such leaks and repair them before the fluid leaks away and a radiator can fail.

An outpost on the Moon requires a continuous supply of consumables, particularly those needed to support the human crew. A typical individual requires daily about 1.84 pounds (0.83 kg) of oxygen, 5.18 pounds (2.35 kg) of drinking water, and 2.6 pounds (1.2 kg) of solid/liquid food. In addition each person must be supported with about 12 pounds (5.4 kg) of water for personal hygiene, and another 1.5 pounds (0.7 kg) of water for preparing food. A special module, referred to as a logistic supply module by NASA, must be designed to satisfy these personal needs. Sufficient resources for an outpost crew to remain on the Moon for six months can be carried to the Moon in a logistics module that can be landed complete with 7,000 pounds (3,000 kg) of liquids (e.g., water) and 24,000 pounds (11,000 kg) of solids (e.g., food) by an unmanned cargo-carrying spacecraft. It would be unloaded from the space freighter and transported to the habitat where it would be connected to the logistics interface module. Until fully regenerative systems and food producing systems can be made available to the outpost, one of these logistics modules would have to be sent to the Moon twice a year to support the crews at the outpost.

With the outpost established and fully operational, the exploration of the Moon can begin in earnest. As it proceeds, the outpost will continue to evolve and will be developed into a fully fledged lunar base or system of bases. The lunar base will become actively engaged in a wide range of exploratory work and scientific investigations covering many disciplines and reaping for humanity many benefits derived from the initial explorations of the Apollo astronauts. This activity is discussed in more detail in the next chapter.

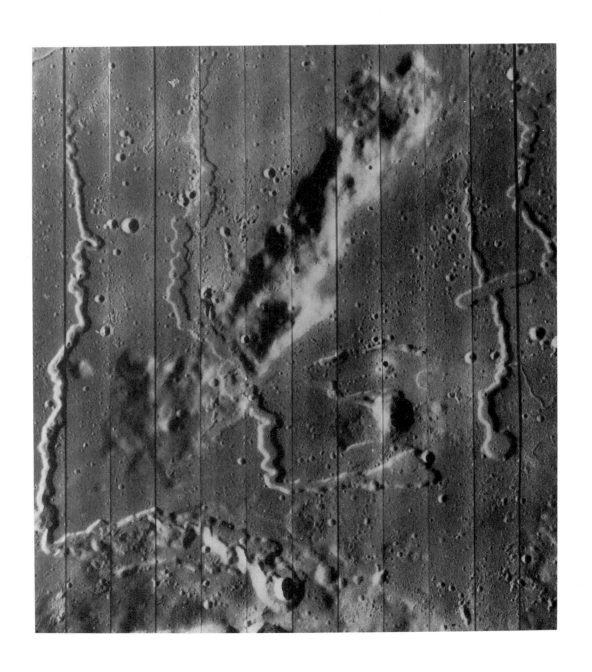

7 FROM OUTPOST TO BASES

As stressed several times in earlier chapters, we cannot effectively explore the Solar System by machines alone. While machines can perform the initial pioneering sorties, humans, with their intuition and flexibility, are needed for in-depth exploration. The Viking expedition to Mars is a good example of the failure of machines to come up with answers. The science investigation was based on the best information available for the design of equipment to seek evidence of life. The results were inconclusive because there were unexpected data. A person stationed on Mars could have modified the experiments to clarify the data and would have been more capable of obtaining answers. With Apollo the astronauts were able to investigate targets of opportunity that most probably would have been missed by automated rovers, and in the later missions, a scientist on the Moon was able to make extremely valuable contributions that would not have been possible without his background and training.

Because the human element is a vital aspect of humankind's expansion into the Solar System, protecting and sustaining the lives of the human explorers is mandatory. Yet at the same time we cannot hesitate in fear of losses. Human progress has been marked by many daring individuals who were willing to risk their lives to explore the Earth and to develop machines to give humanity a greater capability, machines to carry explorers across the oceans and into the air. Without these risk-taking explorers we would not have ventured across the oceans of Earth, descended into the depths of the oceans, and mastered the ability to fly through Earth's atmosphere. As humans push into space there will undoubtedly be accidents and losses. But this does not mean we should not make the attempt. Let us not behave like acorns scared to germinate and grow into oak trees because they were fearful that some might be felled for furniture.

If we are to embark upon the great adventure of expansion into space we must accommodate the needs of men and women in hostile circumstances quite different from, and far away from, terrestrial environments. The nearest terrestrial analogy is the long voyages of the early navigators who left the security of Europe to roam the oceans in hostile environments. During these voyages the explorers were separated from civilization for years. For many of those pushing to Earth's far frontiers there were voyages of no return. Surely we today are no less adventurous than our forebears! But with a whole universe awaiting our exploration we seem not to have learned the lessons of history. We continue to glorify military activities that rarely serve any useful purpose except to boost our national egos. Nations and ethnic groups continue to make great efforts to lay claims to territories that will inevitably be shattered and reassembled as the slag of continents jostle over the hot magma seething beneath them. Rarely do we settle the economic, sociological, and political problems that cause these conflicts. We seem reluctant to spend equivalent amounts of effort to continue our heritage of exploration into missions beyond the womb of Earth. Indeed, interplanetary humans appear to be having a very difficult birth.

There are major fundamental differences between terrestrial exploration and the exploration beyond Earth: changes in gravitational force down to microgravity, inadequate atmospheres to support terrestrial life forms including humans, extremes of temperature ranging from deep cold to blazing hot, the speeding bullets of debris from the formation of the Solar System, and hazards of ionizing and destructive radiation from the cosmic background and explosions on the Sun of almost unimaginable violence. Protecting, nurturing, and sustaining those intrepid crew members of our spaceships exploring the Solar System presents a technical challenge of enormous magnitude (figure 7.1). Yet in perspective it seems no more challenging than the design and construction of a jumbo passenger jet might have appeared to the Wright brothers early in the twentieth century had they accepted such a challenge to their ingenuity. Certainly the idea would have been rejected by their contemporaries, just as many of our contemporaries reject the idea of a human role in space.

It is essential, nevertheless, that a good understanding of the physiological and physical effects of flight through space, and of living for extended periods on other worlds, must be gained before human space exploration can flourish. This requires both ground-based and flight research. The environment to be experienced must be simulated, and countermeasures to adverse effects must be developed. While ground-based simulators with people living in closed environments for many months will provide some answers, space-based simulators are most important. The space station Freedom is an essential research tool for the new assault on the Moon and the establishment of a permanent base on our satellite. In turn, experience on the Moon provides the necessary stepping stone to a more distant permanent establishment on Mars, followed by activities on asteroids and satellites of other planets, ultimately leading to missions beyond the Solar System and our evolution to interstellar species of more intelligent and hopefully wiser beings.

FIGURE 7.1 Establishing a permanent human presence on other worlds presents a technical challenge of enormous magnitude to protect, nurture, and sustain people in routine flights through space and in their operations in inimical environments on the Moon and Mars. This artwork shows a member of the initial base crew operating outside the habitat as a research cart is being prepared for gathering data. (*NASA-Johnson*)

• RADIATION HAZARDS

There are basic problems that have to be solved if humans are to operate on other worlds. For example, in connection with human operations on the Moon explorers must be protected from solar storms. The magnetic field of the Earth acts as a widespread shield close to and on the surface of our planet. It acts as a barrier to particles emitted by solar flares and to most galactic cosmic ray particles. Astronauts in LEO, such as those in Russian space stations and the American Space Shuttle Orbiter, are protected by this field. Unfortunately the same cannot be said for astronauts journeying to the Moon or for those unprotected on the lunar surface.

How this exposure to interplanetary radiation will affect living cells is still not understood; most experiments so far have been made toward an understanding of the radiation effects to be experienced in LEO, particularly with a view to safeguarding astronauts who will spend longer periods in space when space station Freedom is operational.

Without some means of protection it is possible that crew members on journeys to the Moon and during long-term operations on the lunar surface outside a lunar habitat—for example, long exploratory traverses in an unprotected vehicle—could experience radiation exposures exceeding limits now established for astronauts in

LEO. Without adequate protection for the astronauts, a solar flare could be devastating, leading to serious radiation damage, radiation sickness, and even death. While there are reasonable methods to protect astronauts from solar flare particles, high-energy cosmic ray particles present a more serious challenge. To screen from such particles requires massive absorbers or strong magnetic fields. The long-term chronic effects of the galactic cosmic rays are unknown; they may include damage to genetic material of cells, formation of cataracts, and even triggering of fatal malignancies.

While solar flare particles can be shielded against, it may be impractical to shield against cosmic ray particles. To gain some data on the cosmic ray hazard, NASA proposes launching satellites into polar orbit. Cosmic rays can penetrate earthward more easily at the magnetic poles where the flux lines are vertical and the charged cosmic ray particles can travel along the flux lines instead of being deflected when they attempt to pass across a flux line. Plants, rodents, other small organisms, and cell and tissue culture will be exposed to the cosmic rays and the effects monitored.

The best solution appears to be use of the propellant or cargo of materials such as water to absorb some of the cosmic rays. But the high-energy particles produce secondary particles of less energy which may, in fact, be more biologically damaging than the primary dose. These effects must be studied and evaluated in detail. Fortunately, when the astronauts are on the surface of the Moon, its bulk will shield them from half the cosmic ray flux and half the exposure. Nevertheless, if the consequences of exposure to cosmic rays are insurmountable, this flux may alone confine humans to Earth and limit their expansion into the Solar System, acting as a planetary quarantine until humans are much more knowledgeable than they are today.

The acute radiation hazard will, of course, be the particles sent hurtling across interplanetary space by solar flares, those energetic outbursts of energy with the power of billions of nuclear weapons. This radiation can be so intense that it could lead to the swift death of an unprotected astronaut beyond Earth's magnetic field. Fortunately, protection is relatively easy compared with any attempt to protect from galactic cosmic particles. Solar storm shelters can be constructed in those parts of the spacecraft and the lunar base habitats that are already shielded by structure or cargo. Studies are now needed to find the best materials to use in the walls of the shelters to provide additional protection. Strong magnetic anomalies, if discovered, might help toward achieving this protection at specific locations on the lunar surface.

Returning to the question of hazards from solar flares, an adequate warning system rather than a shielding mechanism is the technical challenge with regard to solar particles. Fortunately for the lunar crews the problem is somewhat less than that for crews on the surface of Mars. Astronomers keep a constant watch on the Sun from Earth and quickly detect the occurrence of flares around the solar limb and on the Earth-facing hemisphere. For astronauts going to and on Mars the problem is compounded by the fact that the Mars-bound spacecraft and Mars itself will sometimes be on the far side of the Sun and liable to be affected by flares that are hidden from Earth-based observations. Moreover communication distances are greater than for the Moon, thus eroding some of the warning time.

Extravehicular activity on the Moon must not be scheduled when flare activity is expected. If a flare erupts unexpectedly, the personnel on the Moon, or en route to it, must be warned in adequate time for them to move to a storm shelter before the particles from the flare reach the spacecraft or the Moon. The unpredictability of these events results in warning times of only 30 minutes or less. Already NASA has proposed several approaches to safeguarding the lunar astronauts, approaches that will have later payoffs during missions to establish a permanent human presence on Mars.

Stepped up observations of the Sun from Earth, especially research to understand the changing magnetic field of the Sun and its relation to occurrence of flares, will, it is anticipated, lead to a better average in predicting the occurrence of dangerous flares. An Earth-orbiting solar laboratory could be equipped with a solar telescope to monitor the onset of flares continuously. As data were acquired by this laboratory, a solar monitor could be designed and sent into space to orbit with the Earth around the Sun by being at a Sun-Earth libration point. This monitor would be always looking at the Sun, compared with an orbiter that would sometimes be screened from the Sun by the bulk of the Earth. The monitor would automatically emit warnings to lunar transfer vehicles and to the lunar base. While a base on the Earth-facing side of the Moon would be screened from the monitor when the Moon is around New, this would not be important because then also the bulk of the Moon would screen the base from solar particles.

Experience with the monitor would prove invaluable when planning missions to Mars because for such missions several monitoring stations will be necessary to ensure that all hazardous flares are detected and warnings issued as required. This warning capability must be assured irrespective of the relative positions of Mars, Earth, and the crew-carrying Mars–Earth transfer spacecraft.

• HUMAN COMFORT AND COOPERATION

To operate effectively in space and when on the surface of the Moon crew members must be psychologically and physiologically comfortable. Ideally they should not merely survive in these alien environments but also be able to thrive there and complete many demanding tasks that will be placed upon them. They must operate harmoniously as teams with members of several engineering and scientific disciples, both sexes, several religious beliefs, several races, different national and political affiliations, and diverse backgrounds and goals. The best analogy has been the bases on the Antarctic continent; another, not so close but representing several conditions, were the oil cities of the North Slope of Alaska.

Health maintenance systems will be required to provide medical care, to monitor and regulate the environment, and to encourage peak human performance. Emphasis must be placed on ensuring self-sufficiency and an ability to meet developing problems under conditions of being isolated from Earth and its facilities. However, as contrasted sharply with the earlier terrestrial activities by mariners exploring the oceans of Earth, the space explorers will benefit enormously from our current telecommunications capabilities by which they can always easily seek advice or technical information from Earth's specialists in all applicable fields of expertise.

Where beliefs or other circumstances make it impossible for crew members to rely upon mental control, holistic meditation, metaphysical, or other nonmedical methods to keep their biological systems functioning harmoniously, on-site medical care of the traditional variety must be provided. This care should be sufficient to deal with minor illnesses and injuries and to provide a capability to undertake critical surgical operations, for example in case of an appendicitis, or removal of foreign matter from tissue in the event of an accident.

A fair amount of experience has already been obtained in the use of telesurgery here on Earth. Space-medicine research for Apollo developed many monitoring devices applicable to telemetering the human physical condition for remote specialist analysis or diagnosis over the Earth-Moon distance. In a report from Stanford University in 1987 the example was given of how William Decampli, an astrophysicist who had become a chief surgical resident at Stanford Hospital practiced telesurgery during an incident a thousand miles off the coast of California. A fire aboard a tugboat had seriously injured four sailors. Decampli flew out across the Pacific Ocean to the tugboat and while his airplane circled overhead he radioed instructions to paramedics performing emergency procedures on the injured men.

By the time a lunar base is established it is fairly certain that telesurgery will be commonplace in view of current developments in virtual reality systems and teleoperators. Virtual reality systems mimic the real world and can provide training for telesurgery. They might even be developed to allow a remote surgeon to conduct microsurgery as though he were a miniaturized being in the operating theater. Teleoperators, will initially be developed to allow a single person to control all manner of robotic operations remotely. They could be used to handle automatically complex items of equipment, to assemble a space station, to erect an antenna on the lunar surface, or to service the "hot" sections of a nuclear reactor. In medicine, in the case of accidents at the lunar base, these robotic machines might be programmed to conduct invasive or reconstructive surgery on humans under the supervision of a remote specialist who would have the ability to override the teleoperator at any time should that be needed. This remote surgery is more feasible for use in space stations and on the Moon, where the radio signal transit time is a few seconds, as contrasted with its use on Mars, where the communication time is in minutes.

Meanwhile, until advanced new telesurgery technologies are developed, the medical system developed for the Shuttle Orbiter provides a unit that could be used in lunar exploration, especially for application in the spacecraft between LEO and LLO and in vehicles operating on the surface. The health maintenance facility to be developed for space station Freedom will be applicable to the lunar outpost for it will provide for diagnosis, monitoring, and transmission of data about patients to Earth-based facilities, as well as treatment of minor medical problems. Both the Shuttle unit and the Freedom unit would require some development work to adapt to the lunar operations and missions. An important capability of the lunar health care unit will be a facility to provide controlled pressures for treatment of the bends; to correct a release of gas bubbles into the blood when a rapid decrease in ambient pressure has occurred. Experience in designing and using such units has been gained during oceanographic research projects.

From the psychological aspect, conditions have to be thoughtfully orchestrated to normalize living as much as possible, especially for crews who will be separated from terrestrial circumstances for many months and will not only experience a different gravity but also be closeted in a confined volume. Experience with nuclear submarine crews will provide valuable data to help preserve wellbeing. From a purely practical standpoint, however, human individuals vary enormously in their tastes and in their beliefs in what is healthy or unhealthy. Anyone who has had the task of managing a diverse group of people in an air-conditioned building knows of the problems of finding an equitable setting for the thermostat. Experience on Earth at remote sites of technological effort has shown that there are at least three distinct groups of people who get involved in the activities of physical exploration: those who thrive on an adventurous life the challenges of which they enjoy accepting and overcoming, those who have a romantic notion about adventure but find that they cannot accept the challenges and want to get back to normal as soon as possible, and those who hate the challenging environment but go there for the earning potential to amass as much wealth as possible in terms of money or technical qualifications before getting back to a normal environment.

In any group of individuals those who like the adventure and thrive on it are a definite minority. So in selecting crews for the lunar mission it is important to choose people of this minority type rather than offer large financial incentives or encourage romantic dreamers who want to escape from realities of Earth to the fantasy of an ideal world on the Moon.

If establishing the lunar outpost and base becomes an international rather than an American exploration there will be the problem of selecting crew members of even more diverse viewpoints. An international race or sex quota system must be avoided at all costs. It is mandatory that for the success of the lunar operation, crew members are selected on the basis of qualifications, versatility, harmony, and dedication to the human exploration mission: the expansion of human life forms into the Solar System rather than the expansion of specific human systems or ideologies or a balanced mix of ethnic groups into the Solar System.

Another important point has to be remembered concerning an international crew; the technology required on the Moon may not necessarily be that of the so-called advanced industrial nations. The technology of the latter is mainly water intensive, energy intensive, and aimed toward maximizing profit for a few instead of survival for many; it is bound to terrestrial resources and environment and to terrestrial industrial revolution cultures. Other cultures that are not educated in advanced-nation high technology may have equally important contributions to make because they have often had to rely upon their ingenuity to survive and thrive under adverse economic or physical conditions. Such adaptability may be a great advantage in developing nonterrestrial technology for operation in the alien environments of other worlds.

Social organization also may have to be changed as the lunar base expands. Different cultural philosophies have begot different social organizations here on Earth, and even the mounting of the missions to establish a human presence on the Moon may require radical changes to how the federal government operates, changes to the procurement process, the budgeting of costs, and the management of the whole operation. Indeed it was only by abandoning several accepted federal processes that the U.S. was able to catch up with and match the accomplishments of the Soviet Union during the early days of the space program. The successes resulted from the dedication of a few individuals organizing teams to get results where before the typical operations had blundered on with failure after failure. Miracles of management and cost-effective operations pushed a fleet of spacecraft to the ends of the Solar System before the bureaucracy of Washington stifled the space activities and new starts became an oddity while existing space programs had to fight every year for their minuscule budgets. At the same time astronomical sums were wasted in pork-barrel projects and bureaucratic inefficiencies.

Before starting a program to establish a base on the Moon, it is important that the purpose of the base be understood not only by those specialized enthusiasts who will benefit most in the near future politically, scientifically, or economically from the business opportunities that it creates for certain sections of industry, but also for the long-term benefit of humanity as a whole. In addition, a strong case must be made so that taxpayers understand the cost effectiveness of a continued and expanded exploration of the Moon in terms of the whole of society and its aspirations. A momentous and far-reaching decision has to be made in the near future. Is the human species going to remain confined on this planet with all its physical and economic limitations to further human development, or shall we expand into the Solar System with the aim of ultimately going beyond even its confines?

• NEW RESOURCES FOR HUMANKIND

Science research *in situ* by humans, the emplacement of automatic instruments to gather data unattended at remote regions on the lunar surface, and continued interpretation of the scientific data at universities and research establishments here on Earth are worthwhile activities that might in themselves justify the human effort to set up a lunar base. The mining of the Moon and utilization of its materials both there and in space applications has been suggested as a worthwhile goal, especially in connection with the establishment of large human settlements at the Lagrangian equilibrium points in Earth orbit (points at which the gravitational effects of Earth and Moon balance and a body placed there maintains a fixed relation with respect to the Earth and Moon), and in development of radically new propulsion systems in the form of mass movers and the like for exploration of the Solar System. Lunar materials have been suggested as basic material for constructing and shielding large operations. For example, solar-power satellites could provide electrical energy to Earth without degrading the terrestrial environment.

It appears reasonable to assume that lunar materials can be used to construct and protect a lunar base, and they also might be useful for making rocket propellants for use by the lunar program and possibly for use in other space programs, such as the support of Martian bases or the mining of asteroids as was proposed by Dandridge Cole back in the late 1950s. However, it does not appear to be economically viable to export oxygen derived from lunar rocks (figure 7.2) to LEO to support the lunar missions. The cost of establishing mining operations on the Moon will be extremely high, but so was the cost to the Europeans of establishing supporting bases on the newly discovered continents of Earth. Initially it seems that it will be more cost effective to transport oxygen from Earth to LEO, especially when heavy-lift launch vehicles are developed. On the plus side, however, is the fact that importing lunar materials to LEO instead of lifting terrestrial materials to LEO will reduce the amount of rocket exhaust components that are ejected into the terrestrial atmosphere by the fleet of heavy-lift launch vehicles. Nevertheless, despite the environmental impact of transporting material from Earth's surface to LEO, which would, in fact, be unimportant compared with the pollution arising from other human activities on Earth and vanishingly small compared with volcanic pollution of the terrestrial atmosphere, this economic situation will probably last for a considerable time. It will continue until a large lunar base has been developed and mining equipment has been set up on the Moon to process the lunar materials for several purposes, including providing energy sources for transportation on the Moon's surface as well as for lunar structures and road building. Other cost effective uses of lunar oxygen probably will be in providing propellants for transfer between the lunar surface and lunar orbit.

In 1991 a researcher at the University of Arizona, Kumar Ramohalli, a professor of aerospace and mechanical engineering, demonstrated how bricks, beams, and solar collectors can be made from simulated lunar soil. This material was produced by a University of Minnesota researcher and was made from titanium-rich rock

FIGURE 7.2 Full development of extraterrestrial resources will require explorers and settlers to "live-off-the-land." On the Moon there will be need to process lunar materials to produce building materials and consumables. This illustration shows a NASA artist's concept of a lunar oxygen-production plant. Surface mining moves lunar rocks containing oxygen to the processing plant—the tower structure. Oxygen is stored in a tank farm and the byproducts of rocky rubble are used for building materials and for roadbeds. (*NASA-Johnson*)

mined in northern Minnesota. The pulverized rock is nearly identical in chemical composition and particle size to soil samples returned from the Moon.

The small, disk-shaped solar collector that Ramohalli demonstrated has a reflectivity of about 70 percent. "The solar collectors are not high quality, but we don't care," said Ramohalli. "We would rather take something that is lower quality, but completely made locally [on the Moon], than something of high quality from the Earth." While it would cost thousands of dollars to take such a collector to the Moon from Earth, it can easily be made on the Moon from lunar material. "The whole charm of this is that you can make a small solar collector that produces enough energy to mine more material to make a larger collector," he added. "Then the larger one makes even more energy, and the whole process grows almost exponentially. You become completely self-sufficient very quickly."

All the items made up to the time of writing, including structural members, had been made by "rule of thumb" but the researchers had started to investigate the basic properties of the material under a contract from NASA. The basic understanding of material properties will form an important part of the information needed to build a self-sufficient lunar base, which is, of course, a stepping stone toward the human exploration of Mars.

There might even be valuable raw materials discovered on the Moon that have a sufficiently good value-to-weight ratio to make them acceptable as exports to Earth. One such valuable export is helium-3, that is plentiful on the Moon but extremely scarce on Earth and would be invaluable in developing current concepts of fusion power sources for terrestrial applications. At present nuclear physicists striving to develop controlled fusion power to solve the terrestrial energy problems have to use tritium; but this presents problems. The tritium/deuterium reaction now planned for fusion reactors develops neutrons that incite problematical radiation effects in structures. By contrast, a helium-3/deuterium reaction would develop protons instead of neutrons, which could result in an environmentally more acceptable fusion reactor to solve the anticipated serious energy problems of the twenty-first century.

A NASA study concluded that there is sufficient helium-3 on the Moon to support terrestrial fusion reactors for thousands of years and thus ensure virtually pollution-free energy sources for the whole of mankind. However, one has to accept that other methods may be developed that will not require the use of helium-3 to release thermonuclear power for peaceful purposes. We may never need to mine helium-3 on the Moon!

It seems reasonable to assume that in-depth exploration of the Moon will undoubtedly find other important materials that are scarce or unobtainable on Earth but will have tremendous impact on future human developments here. This has been a characteristic of human explorations of terrestrial continents. The unexpected has often been more valuable in the long run than the known riches initially sought after by the explorers.

There have, of course, been workshops sponsored by NASA on the methodology for evaluating potential lunar resources and how they can be utilized. One held in 1978 at the Lunar and Planetary Institute and documented by NASA Johnson Space Center, discussed three scenarios in which activities would parallel the various classes of quarry and mine development on Earth; among them, moving material to make roads, mining the abundant materials, and development of rare and concentrated resources such as metal ores. An initial activity would be to take samples of lunar material and submit it to physical and chemical processes to ascertain what useful products could be made from the material, similar to the work with the solar collector described above.

The report of the workshop (*NASA Tech. Memo. 58235*) recommended:

> A prudent yet aggressive research and development program would include the development of physical and chemical lunar soil simulants; the production of a map of lunar resources covering all the areas of the Moon covered by Apollo photography and incorporating information from the many geological, geochemical, and geophysical data sets now available; the development of robotic systems with emphasis on increased autonomy of the system; and development of a processing technology so that a technologically reasonable choice of utilization scenario could be made.

In analyzing the use of lunar materials either on the Moon to construct and operate the base or exported to orbit around Earth for use there in, for example, solar power satellites or very large space stations, it will be necessary first to define what materials are needed for large-scale activities on the Moon or in Earth orbit. Then it must be decided which lunar materials are available to meet the needs and what processes are required to convert the lunar materials into a usable form. Finally, the economics of using the lunar materials in these ways must be compared with other means of meeting equivalent objectives.

It is important to understand that only a very small part of the Moon has been explored by the Apollo missions. There may be many surprises (figures 7.3 and 7.4) when an in-depth exploration takes place. The easiest material to obtain from the lunar surface is glass. Aluminum can be obtained from the highland rocks if the mining site is carefully selected to be in an area where the amount of aluminum is relatively high. Iron and titanium can be extracted from the mare regions where there are concentrations of ilmenite. One of the Russian landers discovered a site where the rocks were enriched with iron compared with titanium. Oxygen, too, can be extracted from the ilmenite.

The most easily obtained bulk material will be through mining the regolith rather than attempting to deep mine the underlying hard rock, at least in the beginning of human activities on the Moon. Mining activities might use conveyors or trucks to carry the bulk material to the beneficiation sites. And the whole mining process could be automated with a minimum of human control, which could be remote rather than at the mining site. A mining site should be where the regolith is relatively fine-grained without many large rocks. To extract desired materials from the bulk material after mining, the important factor on the Moon is the scarcity of water. So electrical rather than fluid processing will be required. Finally, refining can be electrical or thermal, both of which sources of energy can be obtained on the Moon from solar radiation.

A significant result of the early studies (1980s) was that at least some lunar resources could be extracted and converted to useful products on the Moon without requiring any major development in technology.

FIGURE 7.3 There are many odd features on the Moon and certainly surprises must be expected when an in-depth exploration takes place. This Lunar Orbiter image shows a part of the lunar surface within a crater upland basin where there are unusual large protuberances in an area measuring 750 by 550 feet (228 by 168 m). The largest of these is 50 feet (15 m) wide at the base and about 70 feet (21 m) high. (*Boeing Company*)

• EXPANSION TO A BASE

When sufficient experience has been gained with the outpost, and several crews have been rotated to prove the operational reliability of the various space transportation systems and of the use of aeroassist to conserve propellants on return to LEO, an expanded habitat will be needed. This will accommodate larger crews and allow for longer stays. It will also provide a larger pressurized volume for operations, logistics, and science.

As was described in the previous chapter the initial outpost will consist of a constructible 50-foot (15-m) diameter sphere located either in a suitable crater or an excavated hole in the lunar surface (refer to figure 6.35). The sphere will be partially buried, probably to a depth approaching 30 feet (10 m). To provide an equivalent

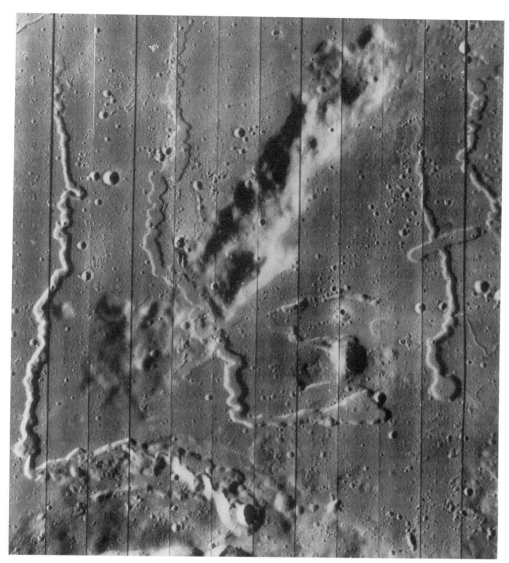

FIGURE 7.4 This Lunar Orbiter photograph shows an intricate group of valleys in the Prinz region of the Moon's Harbinger Mountains located at 43 degrees West longitude and 27 degrees North latitude. At the left is Rima Prinz II which starts at a 2-mile (3.2-km) diameter crater at the bottom of the picture, makes a right-angled turn and runs north for about 40 miles (65 km). The valley is about 1.5 miles (2.4 km) wide to begin with but narrows to about 1,500 feet (460 m) near its end. Lying inside the broad valley is a smaller channel less than 1000 feet (300 m) wide. Rima Prinz II drops about 4,000 feet (1,200 m) in elevation from its beginning on the flanks of Prinz crater (bottom edge of the picture) to its end on the mare surface. It appears to be a record of two stages of volcanics: an older broad-valley producing stage and a later stage of lesser intensity in which fluidized lava cut the smaller channels within the larger. Areas such as this are candidates for exploration to seek lunar volatiles. (*NASA*)

volume, prefabricated modules assembled into a complex would require at least ten times the mass of the constructible habitat. The main spherical structure is assembled by pressurization, and then within it an internal load-supported framework is raised on telescoping columns to provide several floor levels and to support each floor. A construction shack provides accommodation for the crews while they are working to assemble the spherical habitat, which, once it has been pressurized, is connected by a pressurized tunnel to the construction shack. Outfitting of the interior can then proceed in a shirtsleeves environment. The levels within the sphere have different functions. Topmost is the crew support and living area. Below that a floor contains all the equipment and consoles for base logistics and operations and mission operations. Another level down contains the crew quarters, while the lowest level contains the environment control and life-support system.

From the mission operations level a tunnel leads through an airlock to the lunar surface. Another tunnel connects to the construction shack as mentioned earlier. The habitat sphere is surfaced with a thermal coating to minimize heat transfer. Once the sphere is in place, powered mobile equipment will be used to fill plastic bags like sausage skins with lunar regolith. The equipment will lay this long "sandbag" in a coil around the habitat to provide protection from the lunar environment (figure 7.5).

As the base develops similar installations will be placed elsewhere on the Moon, and at the most important locations several habitat structures will be clustered

FIGURE 7.5 An operating lunar base developed from the initial outpost is shown in this artist's concept from a Johnson Space Center study report. In the distance a surface vehicle transport is bringing cargo and personnel along the gravel road from the spaceport to the habitat area. The inflated habitat is in the process of being covered with regolith by a machine just behind and to its right. Space-suited workers are putting finishing touches to the protective covering of the construction shack, and other workers in the foreground are testing an experimental six-legged walker which may have advantages over wheeled vehicles for explorations in rough surface areas. In the right background is the solar power system. Over to the left in the distance are the solar power collectors for the oxygen production plant. (*NASA-Johnson*)

together to form lunar "villages." Ultimately, however, exploration is expected to lead to discovery of underground caverns, old lava tubes, which can be sealed and pressurized and within which can be installed facilities for large lunar bases housing many people.

An environment control and life-support system is a vital element of any lunar base system. To begin with, for the early outposts, this system will be simple and will rely much on supply of consumables from Earth. Later, of course, as the base is expanded and more people are located on the Moon at various facilities and for diverse operations and missions, regeneration and manufacture of life-support consumables will be necessary. Essentially an effective system will have several clearly defined characteristics. First, it must be compatible with the concept of a gradual, modular expansion of human presence on the Moon consistent with addition of extra modules. It must provide identifiable logistics benefits for both short- and long-term missions, particularly in being independent of Earth supplies. Also it must be easily installed at sites on the Moon, and it must be extremely reliable without demanding much time or effort of personnel on the Moon for its operation and maintenance. There must be extra components and alternate subsystems within the module to permit the system to be reconditioned, repaired, or its components replaced, without interrupting normal operations of the lunar base crews. Such upkeep and repair should be possible to the crew of the base without requiring technicians from Earth.

However, it is expected that there will be some advantages by having completely automatic operation of the environment control and life-support system with automatic maintenance and computer-generated alerts both to lunar base personnel and to monitors back on Earth. Ability of some control and maintenance by remote control from Earth-based monitors might be necessary to safeguard against emergencies in which the lunar personnel could be incapacitated and at further danger should the life-support system also malfunction; for example, inability of the crew to gain access to the environment control and life-support system module because of an emergency or breakdown of other equipment.

Because individual lunar outposts may not be occupied continuously during the early period of developing a human presence on the Moon, the environment control and life-support system must be capable of being shut down and left unattended for periods and then brought back on line when required to reactivate an outpost or a module. This capability will be needed should the lunar base have to be vacated by its crews sometimes, to be mothballed for a subsequent return of personnel. It would be analogous to the initial exploration of Antarctica where bases were often manned only during certain seasons.

The life-support system will be in several forms: a simple system for use in a single personnel module of a lunar outpost, a specialized central and separate module that relies on Earth to replenish consumables to support a group of other habitable modules, a specialized central and separate module that recycles or regenerates consumables and is nearly independent of Earth supply, an advanced design that not only recycles and regenerates consumables but also generates materials from the lunar resources and is completely independent of Earth supply.

As mentioned in the previous chapter, while initial establishments on the lunar surface might operate through solar power, the long 354-hour lunar night makes insurmountable demands on energy storage. Nuclear power systems are needed to provide the range of power levels (500 kilowatts to 1 megawatt) on a continuous basis. An early study of how nuclear power could be adapted for use of a lunar base was completed by Westinghouse in 1964. The study showed that a compact nuclear plant could be transported to the Moon within the cone-shaped payload envelope of a Saturn V. The plant would produce 100 kilowatts of electricity and could support a complex of auxiliary power supply units for fuel regeneration and oxygen and water supply as well as the habitable modules of the lunar base.

An initial concept by NASA, known as the SP-100, was developed in the 1980s to use dynamic or thermoelectric conversion systems to develop 100 kilowatts. As the base grows larger, units would be needed using dynamic conversion systems to produce at least 500 kilowatts. The reactor core would be buried so that lunar soil would provide shielding. On the surface would be thermal radiators to reject unused heat, and the power conversion and conditioning unit. The reactor unit must have low specific mass. Also the system must be somewhat easy to bring to operational status, say within a few days, and once it becomes operational it should be extremely reliable with a need for very little support by the base crew. That nuclear power systems can achieve such high reliability has been demonstrated in the management and organization of the French nuclear energy program as contrasted with other countries where nuclear power generating systems have suffered various unacceptable levels of accidents and system downtime. Systems have also been demonstrated and operated safely for many years in naval vessels.

Many important future uses of the Moon can be visualized, but international treaties will have to be changed. In the past there were fears that the Moon might be used for military purposes or for the sole benefit of aggressive nations. In 1979 the United Nations Committee on the Peaceful Uses of Outer Space drafted a treaty intended to regulate the expansion of humans into space, especially to protect the rights of nations without current space-faring capabilities as a bulwark against the typical imperialism that had marked exploration and exploitation of the Earth. Consequently an international Moon treaty placed severe restrictions on what might be done on the Moon, banning ownership of resources on the Moon but permitting their use. Modeled after earlier treaties to protect exploitation of the Antarctic Continent, the treaty now needs revision so there can be no question about allowing use of the Moon to establish bases there and to utilize its resources in many ways, not the least of which might be to support Earth satellites, satellite power stations, and even space settlements as visualized by Gerard K. O'Neill in his book *The High Frontier* (New York: William Morrow, 1977) and the NASA-Ames Summer Studies of 1975 (*Space Settlements,* NASA SP-413, Washington, D.C., 1977) and 1977 (*Space Resources and Space Settlements,* NASA SP-428, Washington, D.C.,1979).

Without changes to the treaty controlling use of the Moon, there might be no encouragement for private enterprise to develop the lunar resources if every recovered resource must be shared equally with all the peoples of Earth, even those who could not at present care less about developing technology to establish bases on the Moon, or be subjected to some unexpected frivolous legal challenge. While the idea of establishing the Moon as a new economic commons, the property of all humankind and controlled by the United Nations, might appear to some as a resurrection of the ideas of Marx and Lenin, proponents of the treaty claimed that it would encourage private enterprise on the Moon by ensuring a political stability there by avoiding any clashes arising from national ownership. While this may be true, much has changed in the world since 1979, politically, economically, and technologically. These changes would suggest that international treaties concerning how humankind moves from Earth into the Solar System to develop its resources, and particularly that first stage of establishing a permanent presence on the Moon, should be carefully scrutinized, not only by legal 'experts' and sociologists but also by entrepreneurs and visionaries and, if necessary, redrafted. We have all seen how constitutions and methods of government that were established in the eighteenth century are leading to serious problems in the late twentieth century. Human beings just do not appear to be smart or wise enough to look very far ahead and establish rules and regulations that will endure as society, mores, politics, and technology inevitably change.

At the final press briefing following the 1977 summer study O'Neill described how mass drivers—electrically powered machines that accelerate materials to high speed and send them out in a precise direction, analogous to a rocket engine—would consist of a long "guideway" made of aluminum strips, wound with coils. A small vehicle with no moving parts, referred to as a bucket, floats along the guideway by magnetic forces. It accelerates to high speed, pushed by the magnetic forces, then releases its payload and returns for reuse. Such mass drivers, using the mass of otherwise unusable materials from the Moon, could be applied as launchers for lunar material into space, said O'Neill. They could be used instead of rocket engines to propel vehicles in space to move, for example, heavy loads from LEO to the vicinity of the Moon. Further downstream the drivers might be used to drag asteroids into new orbits where their materials could be more easily mined. Even more important would be use of mass drivers in connection with the survival of the human race and the protection of the terrestrial environment.

There is a group of asteroids, known as the Apollo-Amor group, whose orbits cross or pass close to the orbit of the Earth. About fifty are known but there are probably several hundred thousand presenting a real threat to the continuance of life on Earth. Protecting the planet would rely upon developing an ability to use a mass driver to move any asteroid from an orbit that would ultimately cause a collision with Earth. Thus the establishment of a lunar base might prevent the otherwise inescapable and inevitable widespread destruction of Earth's environment and almost all life forms, including humans still living on the planet. Estimates have been made that such encounters with asteroidal-sized bodies (less than one mile to many miles in diameter) may occur on the average every 300,000 years and that these can vary from being local disasters affecting thousands of square miles to global Armageddon.

It occurs to this author that a lunar base might become much more important in safeguarding terrestrial life forms than current attempts by ecologists here on Earth, where fossildom of species almost seems inevitable. A base could act as a technological Noah's Ark in which genetic material for all known species, including spores, seeds, sperms, and eggs, would be stored on the Moon and later used to restore a biosphere on Earth following a cataclysmic encounter of our planet with an asteroid. Sufficient time would, of course, have to elapse after the collision for the environment to stabilize.

• MINING, INDUSTRY, AND TOURISM

The summer studies also considered that the scenario for mining on the lunar surface would involve automated mining vehicles and conveyor systems rather than trucks. These two methods were mentioned earlier in this chapter. The amount of material to be mined could best be handled by a conveyor system, probably using bucket-wheel excavators. The summer study believed that a mining capacity exceeding 1,000 tons per hour for a maximum yearly output of 600,000 tons of beneficiated ore would be needed by the fifth year of operations on the Moon to support the development of major facilities in LEO—space settlements and solar power systems. Beneficiation reduces the ores so that the mass reaching the refining process does not include much inert material. The recommended beneficiation process was based on electrostatic and magnetic separation. The extraction and construction working group concluded that about 90 pounds (40 kg) of volatiles (60 percent water) could be reclaimed by thermally processing smaller than 20 micron ilmenite concentrate extracted from 130 tons of ilmenite ore. This seems rather optimistic unless water-bearing rocks are discovered in the shadowed regions around the lunar poles. Fortunately the lunar materials do not need crushing machinery because they are fine grained to begin with. Aluminum could be extracted from plagioclase concentrate and titanium and iron from ilmenite by carbochlorination methods. Carbochlorination to extract titanium and iron from ilmenite, aluminum from plagioclase enriched with anorthite, and magnesium from olivine, uses chlorine at low temperatures. When aluminum is the end product, additional processing by electrolysis will be required. For obtaining titanium, additional reduction by calcium followed by hydrogen will be needed, and for iron the hydrogen reduction will suffice.

Oxygen and carbon dioxide might be recovered from ore concentrates as byproducts of the metal extraction processes. Fiberglass can be made from lunar soil, with inorganic adhesives made from the lunar soil used to bind the fiberglass into usable forms. Such adhesives will be needed also to form paving or building materials directly from lunar soil and from slag resulting from processing of the regolith to extract minerals. A solar furnace could be used to provide heat for manufacturing the fiberglass. Raw materials to make the fiberglass would be mainly (80 percent) slag derived from the production of aluminum from plagioclase, and small percentages of calcium oxide, a byproduct of the production of titanium, and anorthite. Ground into fine grains the material would be melted and poured into molds. It could then be used either in block form for building or extruded into filaments for use in filament-wound structures or for weaving into textiles.

The metal extraction processes will have large amounts of carbon monoxide as a byproduct. This can be used to produce oxygen while recovering the solid carbon for recycling to the metal extraction process. A process used in the petrochemical industry for many years makes use of carbon monoxide reduced with hydrogen catalytically to produce water and methane, both of which will be in demand at the lunar base.

The unique environment of the Moon offers opportunities to develop industries there such as those requiring use of high vacuum technology. It was suggested by G. Harry Stine in his book *The Third Industrial Revolution* (New York: Putnam,

1975) that we could use the resources of the Moon to enter a new industrial revolution based on the Moon where industrial enterprises would be free from the environmental problems of heavy manufacturing. The Moon has some 14 million square miles of undeveloped territory that offer the use of vacuum technology, location for "dirty" manufacturing processes that can no longer be allowed on Earth, and the mining of minerals that are scarce on Earth. However, possible pollution of the lunar environment must be weighed against the value of other uses of the Moon such as for astronomical research.

While the cost of transporting products from Earth to the Moon is excessive, the reverse is not true. The absence of an appreciable atmosphere on the Moon allows uses of electromagnetic launchers that can accelerate payloads to the 1.46 miles per second (2.35 kps) needed to escape lunar gravity. This was first proposed by Arthur C. Clarke ("Electromagnetic Launching as a Major Contribution to Space-Flight," *Journal of the British Interplanetary Society,* 9 (6) (November 1950):261–267) and further developed by this author in a classified report to the Air Force in 1962 in connection with using the Moon as a secure platform for launching military surveillance satellites. Even without the development of mass drivers, such payloads can be projected into the vicinity of the Earth and captured by aerobraking into LEO with a minimum need for rocket propellants. Energy to launch from the Moon can be obtained from the conversion of solar radiation to electrical energy that, in turn, would be used to accelerate the payloads magnetically along the launching rails. Such facilities would be placed in limb regions of the Moon to direct their cargo carriers into the vicinity of Earth. If we can utilize minerals on the Moon and manufacture products there, these products can be sent to Earth, or at least to LEO, without any serious penalty of transportation costs.

It appears certain that the Moon can be the source of many raw materials needed to support industries away from the Earth, and use of lunar resources can help preserve the terrestrial environment while bringing an increased standard of living on a global basis. Moreover, if the lunar resources are used to fabricate solar power stations in orbit around Earth, many terrestrial power problems might be overcome. Space-based power systems would not present the problem of emitting waste heat into the biosphere, nor would they contribute to atmospheric pollution or produce problems of disposal of nuclear wastes. While it may be economically unfeasible, and environmentally unacceptable, to fabricate the space power systems with terrestrial materials and transport the prefabricated parts into LEO or other orbit for assembly there, it would be economically and environmentally acceptable to produce the parts on the Moon and assemble them in Earth orbit. This procedure would avoid any interaction with the terrestrial environment when building the space power system, in addition to safeguarding the environment during its subsequent operation.

It should not be forgotten, either, that in those manufacturing operations on the Moon that rely for energy on solar collectors, the problems of the long lunar night can be overcome by positioning large mirrors in lunar orbit. These mirrors can be fabricated from lunar materials and used to focus sunlight onto the dark hemisphere of the Moon where needed so that solar power stations can operate continuously on the lunar surface if required to do so. This also would mitigate problems of storing large quantities of electricity for night time usage or, if deemed advisable, avoid the use of nuclear power stations on the Moon. The orbiting solar mirrors could also provide nighttime illumination at the base and at mining sites, and reduce the heating requirements of habitats when the sun is below the lunar horizon. By having a series of four mirrors equally spaced in a lunar orbit suitably inclined to the plane of the ecliptic to keep them out of the Moon's shadow, solar radiation could be directed to chosen areas of the lunar surface throughout the lunar night. But interference with optical astronomy would have to be controlled.

When a permanent base is established, those people who now travel on vacations to many obscure or exotic places on the Earth will inevitably wish to travel to the Moon. Its unique recreational and vacation possibilities will undoubtedly be commercially developed. Its unusual scenery, low gravity, and the unforgettable majesty of the blue-green Earth riding in the black lunar sky will be irresistible once sampled. It does not need a great deal of imagination to extrapolate from the sometimes bizarre activities of humans on Earth seeking physical excitement to new types of physical activity on the Moon. Imagine a large hangar on the surface or preferably a cavern within a lava tube filled with an atmosphere with the same pressure as that at Earth's surface. Tourists might be able to don wings and fly using muscle power alone. One can speculate about inventing three-dimensional ball games in which all the team members are equipped with wings. Or we can imagine a trampoline system in which two trampolines are arranged at either side of a large space, each being angled so that gymnasts can bounce from them across the intervening space performing many feats in the low lunar gravity as they almost float from side to side from one trampoline to the other. Might there one day be Lunar Olympics as well as Winter Olympics?

The key to such vacational and recreational development might rest with the discovery and subsequent development of lunar caverns resulting from drained lava tubes at the edges of the big maria (figure 7.6). Such tubes might be effectively sealed from the surface environment and filled with atmosphere generated by processing the lunar regolith and importing some gases from Earth. The underground facilities would have even more important uses than providing an environment for recreation. Suitably equipped with the correct illumination, and provided with water that can be recycled, they would provide facilities for lunar agriculture, a twenty-first century evolution from the systems developed for space station Freedom and the initial lunar base.

• LUNAR AGRICULTURE

Many experiments have been undertaken in recent years by NASA and by university research teams to grow food plants under conditions expected in space stations and on other worlds. For example, the Spring/Summer 1991 issue of *Purdue University News* reported how a team of researchers at the University led by Cary Mitchell was developing methods to sustain life in space. A system referred to as a controlled ecological life-support system integrated plant growth, food processing, and waste management to support life in space and ultimately on other worlds. The aim is to develop a self-supporting system that can operate without importing consumables

FIGURE 7.6   Lava tubes such as this Rima Maskelyn I in the western part of Mare Tranquillitatis may hold the key to future large underground bases on the Moon. In-depth exploration may reveal the presence of enclosed tubes beneath the surface that can be adapted as human habitats. To begin with, self-supporting inflatable habitats could be installed within a lava tube close to where it opens to the surface. Later, sections of tubes could be sealed from the vacuum conditions of the surface and large human settlements established within them, complete with agricultural areas and terrestrial-type buildings. Ultimately a completely independent lunar civilization might develop housed in a network of cavernous lava tubes below the surface and completely independent of Earth. (*NASA*)

from Earth. Plants will provide not just food, but also seeds for later planting, and oxygen for replenishing that used by humans and animals. The plants also use carbon dioxide that would otherwise build up in the atmosphere. All waste would be recycled.

Developing this type of system has been an ongoing project at NASA and universities for several years. Food, water, and a breathable atmosphere are three essentials for humans, and the provision of these essentials on a lunar outpost or base is imperative. Although brief spaceflights can carry some supplies from Earth, long-term activities on other worlds demand that they be produced away from the home planet. The CELSS program of NASA investigates plant growth, food processing, and waste processing and recycling, and the necessary control systems required to automate such an integrated system.

The normal human diet relies upon carbohydrates and fat for its energy source. Edible fats might be synthesized using technology already available. Carbohydrate synthesis is now just beginning to be researched. However, crop plants produce large amounts of carbohydrates by photosynthesis and produce the necessary inedible bulk needed by the human digestive system. Plants, however, require a large volume of growing space and adequate energy sources. The former can be provided by use of lava tube caverns, should they be discovered, and the latter by sunlight channeled to the caverns or by artificial lighting powered by conversion of sunlight at the surface. The great advantage of plants over algae is that they are more palatable to humans, are more visually satisfying, and produce plenty of oxygen to replenish the breathable atmosphere for the human crew at the base.

A plant growth unit was developed during NASA research. For use on the Moon it would be connected to a food processing unit in which the plant material is harvested and converted into prepared foods by microbial, chemical, and enzymatic systems. Edible food products consumed by the crew produce carbon dioxide, water, urine, fecal material, and minor quantities of other waste material (table 7.1). These need to be recycled to make the lunar base self-contained ecologically. Solids and liquid water would be passed through a waste-processing system to produce chemicals that would be used as nutrients for the plants. Waste processing has been investigated in depth; it is a necessary requirement to maintain the health and wellbeing of the human crews on the Moon. Carbon dioxide and water would be recovered and chemically processed for reuse in elementary or molecular form as needed. The plants also produce water vapor that would be extracted and reused either for plant irrigation or purified into potable water for crew members.

Carbon, which is precious on the Moon, would also be recycled. Leaving the plants as biomass (food and inedible plant material), the crew in exhaled air, and as gas resulting from water processing, carbon dioxide would be fed back to the plant growth volume as air enriched with carbon dioxide. This is a stimulant to plant yield and would ensure that carbon was recycled. As the carbon was incorporated in plant structure the oxygen would return to the atmosphere and thus be recycled also.

Experiments conducted at several NASA centers during the 1980s produced encouraging results showing the feasibility and practicality of a controlled ecological life-support system to support long-term missions in space and on other worlds. Plants such as wheat, soybeans, lettuce, and potatoes were studied and produced acceptable harvest yields and time-to-harvest periods. Potatoes supply carbohydrate and soybeans provide fat and protein. The appropriate nutrients for each crop, systems for delivering them, and methods to control bacterial and fungal pests were evaluated. It was ascertained that 99 percent of a human crew's nutritional requirements can be met with cultivation of only three plant types supplemented by yeast and algae to provide nutritional supplements.

At Kennedy Space Center in the late 1980s, a cylindrical biomass production chamber with a volume of 2825 cubic feet (80 cu. m) was used to determine the

TABLE 7.1 *Daily Inputs and Outputs from Human Metabolism*

INPUTS			OUTPUTS		
	lb	**kg**		**lb**	**kg**
Food	1.36	0.62	Solids	0.44	0.20
Water Drinking	5.18	2.35	Water Urine	3.31	1.50
In food	1.59	0.72	Other	4.02	1.82
Oxygen	1.84	0.83	Carbon dioxide	2.20	1.00
TOTAL	9.97	4.52	TOTAL	9.97	4.52

NOTE: Data from NASA Life Sciences Report 1987. Each person will also produce 11,200 Btu of body heat that must be dissipated.

technology requirements for a controlled ecological life-support system module. The steel chamber, which was about the size of a module that could be carried to a space station or to the lunar surface, could be sealed so that the plant cultivation system was completely separated from the terrestrial environment as it would be at a lunar base. Temperature, airflow, humidity, and carbon dioxide levels were controlled, and illumination was provided by lamps simulating about 50 percent of terrestrial sunlight. Plants were first planted in this biomass chamber in 1986, and a full generation of wheat plants was successfully grown to maturity from seed. These plants were harvested.

In fact, by controlling the environmental conditions within the test chamber—a low temperature, an enriched carbon dioxide atmosphere, and a 20/4-hours light-dark cycle, equivalent to full summer daylight—researchers were able to triple wheat crop yield and achieve a fourfold increase over the best terrestrial field agricultural yields. Also, Irish potato, sweet potato, and soybean crops were all tested and grown successfully in test chambers. The yield of potatoes was increased by exposing them to a higher humidity, low temperature, enriched carbon dioxide atmosphere, and special light levels and colors. The rates at which nutrients are needed at different stages of the plants' growth cycles were ascertained—important data for programming an automated plant growth facility.

The effects of increased atmospheric carbon dioxide concentration on plant yield has been known for some time. The author recollects how he interviewed Carl O. Hodge of the University of Arizona's Environmental Research Laboratory in 1969; Hodge had been achieving high yields of vegetables in arid lands close to oceans. While growing the plants in open fields would require use of one thousand times as much water as the plant's final weight, only one-tenth this amount of water was needed in a controlled environment. Moreover Diesel generators used to pump seawater through a system to desalinate the water and regulate the temperature of enclosed plastic greenhouses also provided exhaust gases to enrich the controlled environment in carbon dioxide. Earlier experiments by other researchers had increased yield of rice plants by 100 percent when the normal atmospheric carbon dioxide level of 300 parts per million was increased to 1,200 parts per million. Moreover, the controlled environment, even without carbon dioxide enrichment, produced quick maturity; Bibb lettuce matured in 40 days compared with 60 days in open fields. Yields were remarkably increased; they were five times greater for Bibb lettuce and cucumber, twice as great for peppers and radishes, and 1 1/4 to 1 1/2 times greater for tomatoes and eggplants.

Whereas at the space station soil would have to be imported from Earth, the controlled ecological life-support system facility on the Moon could use lunar soil supplemented with nutrients and minerals brought from Earth or concentrated by processing of the lunar regolith. Terrestrial agricultural experiments showed that arid soil can be used successfully for agricultural purposes. Those by Hodge had demonstrated this in Baja California and elsewhere. A walk across lava flows and cinder cones in places like the Kilauea volcanic area of Hawaii and the Ubehebe crater area of Death Valley, California, will show how flora quickly develop when conditions become ripe in these seemingly inhospitable areas of lava and volcanic cinders. Preliminary studies at Johnson Space Center of agriculture with simulated lunar soils produced very encouraging results.

NASA researchers have concluded that even in the present stage of development of plant growing facilities, enough food could be produced for crew members at a space station or on the Moon by only 130 square feet (12 sq. m) of food production area per person. This would be practical in a space station and would be very easily accomplished within a lunar base, even one that had to be established in enclosures on the lunar surface.

• ASTRONOMY AND OTHER SCIENCE

Although the surface of the Moon presents a harsh environment, inimical to human life unless astronauts are protected against temperature extremes, solar and cosmic radiation, and micrometeoroids, and are provided with a breathable atmosphere at a suitable pressure, that surface may be ideal for some astronomical observations. There are several advantages to locating astronomical observatories on the Moon. The lunar surface is not seriously jarred by moonquakes. A typical moonquake on the average is millions of times less intense than typical earthquakes. There is no lunar atmosphere to distort incoming light from distant stars. No lunar winds or rainstorms exist to shake installations or limit viewing. On the far side of the Moon the light and microwave radiation of Earth is screened by the lunar bulk. Astronomical observatories on the Moon can exceed the resolving power and sensitivity of terrestrial observatories, even the "Great Observatories" being established in LEO, over all regions of the electromagnetic spectrum.

While the Hubble space telescope has the potential of ten times the resolution of ground-based telescopes at visible wavelengths, lunar telescopes could be made with many thousand times an increase in resolution with the ability to find planets around other star systems. A radio telescope on the Moon linked by telecommunications to another on Earth would provide a baseline with the potential of resolving distant radio sources as though it were a giant antenna some 250,000 miles (400,000 km) in diameter. Interferometry uses two or more individual telescopes (optical or radio) linked electronically so that they interfere with each other. The interference patterns can be analyzed mathematically to achieve results equivalent to an enormous telescope whose aperture, and therefore resolving power (ability to separate angularly close objects), is equivalent to the distance between the two separate telescopes. Obviously one telescope on the Moon and another on Earth provides the equivalent of a telescope of mind-boggling size. The potential for new discoveries about the universe is breathtaking.

While the new astronomical observatories being established in low earth orbit will make major contributions in the coming years, they do not have much future potential. To begin with this volume of space close to the Earth is polluted by orbiting debris, Earth-generated radiation, and an extremely rarefied extended atmosphere of dust and gas particles that can be affected by the kinetic energy of the observatories passing through it and by changes in solar radiation causing the extended atmosphere to expand and contract unpredictably. Skylab fell victim to these effects.

Even the Earth's magnetic field presents problems by generating low frequency radio waves that mask a region of the radio spectrum astronomers want to observe. Radio telescopes must be placed beyond the influence of Earth's magnetic field to observe these regions of the radio spectrum. The optimum place is on the Moon.

As astronomical satellites revolve around Earth they pass rapidly from full sunlight into darkness, and the resultant temperature changes present serious stability problems to the optical design. The thermal stresses make it necessary to incorporate complex shields and thermal control systems in the orbiting astronomical observatories. By contrast, on the Moon, while there will be thermal stresses, these vary over the much longer period of the lunation, in days rather than minutes as is experienced in LEO.

In 1965 a NASA summer conference concluded that "the Moon may well offer an attractive and possibly unique base for astronomical observations." Astronomical work that would benefit from location of lunar observatories include radio astronomy, X-ray astronomy, gamma-ray astronomy, optical solar astronomy, and optical planetary and stellar astronomy. Subsequent conferences that planned for the science to be undertaken by Apollo identified some astronomical work that could be performed on the Moon by Apollo astronauts or subsequently when the Apollo program was extended beyond its initial objectives. Unfortunately only very few astronomical experiments were made by Apollo and, even more disappointing, there was no subsequent activity on the Moon.

A series of measurements were defined that would provide data fundamental to establishing an astronomical observatory or observatories on the surface of the Moon. Among these was the flux of micrometeorites that could degrade optical surfaces, the thermal conditions on the lunar surface, the ability of the lunar soil to support instruments firmly, the seismic activity, and the light scattering and obscuring effects of lunar dust. These important data emerged from the Apollo missions and showed that it would be feasible to establish astronomical observatories on the Moon (figure 7.7) that would exceed in their capability anything on Earth or in LEO.

Additionally, observatories demand suitable sites. The early studies ranked this as a priority. Again, the Apollo program showed that there are suitable sites on the Moon where astronomical observatories can be established. Radio, X-ray, and gamma-ray astronomy would require a symmetrical crater some 60 miles (100-km) in diameter with a fairly flat floor and, for X-ray and gamma-ray observations, an even rim rising at least a half mile (about one km) above the floor, while for radio astronomy the site should be free of cliffs within 30 miles (45 km). Also the site should be near the equator to maximize the area of sky that can be covered during each rotation of the Moon on its axis. Optical astronomy would require a site with unobstructed view of the horizon.

All observatories should be on the far side of the Moon to prevent optical and radio interference from Earth. Note that on the Earth facing side the night sky of the Moon would be dominated by a brilliant Earth many times the area of the Moon and, because of its high albedo, many times brighter.

For X-ray and gamma-ray astronomy the Moon offers a superlative base for observations because of its extreme stability and its slow rate of rotation that is accurately known so that positions of stellar objects can be predicted with great precision. Moreover the sharp edges of the Moon's horizon can act as a precise shutter to determine the exact position and angular size of objects occulted as the Moon rotates. This advantage will apply also over a great range of wavelengths. Also, because of the slow movement of the sources across the lunar sky, extremely long exposure times can be given that, coupled with an ability to make detectors of large effective areas, will yield great sensitivity even when observing extremely faint objects. This will push the observation of the sources very much deeper into space. One site suggested in a NASA 1967 Summer Study at Santa Cruz, California (*Lunar Science and Exploration,* NASA SP-157) was the crater Grimaldi on the border of the Oceanus Procellarum. It has the right diameter and a rim of about the right height, and a flat floor with no central mountain. It is close to the lunar equator, at a point where Earth does not dominate the lunar sky. Observed from Grimaldi, the Earth always appears close to the Moon's eastern horizon (figure 7.8).

A site on the Moon has great potential for radio astronomy observations, especially at long wavelengths, where very large arrays are required to obtain adequate resolving power. Arrays several kilometers across cannot practically be erected in

FIGURE 7.7 Ultimately very large optical telescopes might be established on the Moon, benefiting from the low gravity, lack of atmospheric weather, a crystal clear atmosphere, and a seismically quiet surface. Construction workers are here shown mounting the various elements of the huge mirror each of which will be controlled by computer to bring the light rays from distant objects into sharp focus. Observatories on the Moon can have many days of darkness for uninterrupted long-exposure photographs of faint objects. (*JPL*)

FIGURE 7.8 Lunar observatories should be established where interference from Earth will be minimal. A location within the crater Grimaldi, which has a large flat floor without a central mountain, might be suitable for several types of astronomical observations. From that location the Earth would always be close to the lunar horizon, presenting its sequence of monthly phases two of which are shown in these artist's concepts.

Earth orbit because they would be too difficult to stow aboard shuttle craft and to deploy and stabilize in orbit. They might be taken into orbit as prefabricated sections, but this also could be done for transporting them to the Moon. Deploying and assembling these units would be more easily done on the Moon's surface than in LEO, and once set up on the lunar surface, the array would be extremely stable. Also directing an array in LEO would offer almost insurmountable problems compared with directing it from the lunar surface.

Steerable radio telescopes built on Earth suffer from structural problems and distortions of their surfaces by weather conditions and gravity. At one stage of his career with a major aerospace company, the author was involved with researching methods of stabilizing and controlling large radio telescopes and military radar antennas. Although some of these structural and control problems can be compensated for by lasers linked to computers there are limits imposed by the terrestrial environment on the size of a parabolic dish that can be constructed and operated on the Earth's surface. Small antennas are arrayed to provide the effective area of the big antennas that cannot be practical as single dishes. These restrictions could be eased on the airless, lower-gravity Moon. Antennas with much greater gathering power could be constructed and operated, and structural material could ultimately be derived from lunar materials. Some researchers (Ferhat Akgul, Walter H. Gerstle, and Stewart W. Johnson) have proposed building a large radio telescope on the Moon from advanced graphite epoxy materials (*Scientific American,* March 1990). These materials would produce small thermal stresses over the large range of temperatures experienced during the lunation cycle.

Telescopes similar to the radio dish at Arecibo, Puerto Rico might be constructed in lunar craters with diameters much greater than those possible using natural features on Earth. At Arecibo a crater-like depression, formed in limestone by a sink hole, is used to house and support a 1000-foot (300-m) diameter antenna. The antenna beam is moved by supporting a receiver on a movable cradle suspended by cables over the central part of the bowl-shaped antenna. Again, the low lunar gravity would be extremely advantageous in constructing such an antenna on the Moon (figure 7.9). Its diameter might be as large as 5000 feet (1500 m).

Paul N. Swanson, astrophysics program manager at the Jet Propulsion Laboratory, pointed out in a paper to the Space '92 Conference that the first astronomical telescopes may be placed on the Moon early in the return missions and probably before the manned landings. Optical telescopes small enough to be carried to the Moon in automated spacecraft could be automatically deployed on the lunar surface and operated remotely from Earth. He suggested three types of such telescopes: a zenith pointed, wide angle, transit telescope; a student telescope to be operated from any secondary school via a computer and modem, and an automatic, fully steerable telescope of up to 3-feet (1-m) aperture.

Swanson also suggests that when a lunar outpost has been established and manned, interferometers can be constructed on the lunar surface. An initial configuration might have two or three elements only, to be expanded subsequently as cargo flights to the Moon become regular. Three versions of interferometers for use on

FIGURE 7.9   Radio astronomy would benefit enormously from a large dish on the far side of the Moon of the type shown in this artist's concept. (*NASA-Johnson*)

OPTICAL INTERFEROMETER

FIGURE 7.10 An optical interferometer might be set up on the lunar surface consisting of several telescopes (three in this artist's concept) combining their signals in the central facility. Optical path lengths, which have to be precisely maintained, would be controlled by small carts with mirrors on the optical paths. (*JPL*)

the Moon have been studied at the Laboratory. Most ambitious of these is an optical interferometer (figure 7.10) with a baseline of up to three miles (5 km) consisting initially of three optical elements of 5-feet (1.5-m) aperture arranged in a Y-shaped configuration with a central unit to combine the beams. Later, four more optical telescopes would be added to complete the array. To combine the light from all the optical elements and form an image at the central unit, the light path lengths between the telescopes and the central unit have to be equal and controlled with an accuracy of less than one micron. This will require firm foundations and high precision optical mounts for the various elements of the system, followed by monitoring of the path lengths by laser instruments and control of these lengths by moving a small vehicle carrying one of the mirrors in each path from a telescope to the central processing unit. The firm foundation requirement is eased on the Moon because of absence of groundwater and seismic disturbances and the insulating nature of the regolith preventing temperature effects although lunar surface temperatures vary about 300 degrees during the lunar day/night cycle. The regolith might be consolidated for a foundation by sintering it using concentrated solar heat to melt the glasses in the regolith and bind the particles firmly together.

A radio interferometer, working at submillimeter wavelengths (figure 7.11) would consist of seven antennas each 16 feet (5 m) in diameter arranged as a Y with a baseline of 0.6 miles (1 km). Although operating on the same principles as the optical interferometer, this radio system would not have to be so precisely located because it operates at longer wavelengths. However a good mechanical and thermally stable connection to a consolidated regolith is still needed. The larger size of the telescopes will, however, mean that more effort will be needed to prepare the sites. Each radio telescope should be housed in a tent or hut to allow for control of temperature, and the receiving detector should be cryogenically cooled to achieve high sensitivity. The system allows heterodyne detection so that the signals can be

FIGURE 7.11 A submillimeter aperture synthesis array, shown in this drawing as consisting of six elements in a Y-shape would be used to detect radio signals not possible from Earth' surface or impractical from Earth orbit. (*JPL*)

mixed at an intermediate frequency and extremely precise measurement and control of the path lengths from the antennas to the combining unit is unnecessary.

Also at radio frequencies a very low frequency array (figure 7.12) would operate below 30 mHz and have up to 100 elements arranged along a 125-mile (200-km) baseline. Such an instrument, located on the lunar far side would be used to explore the cosmos in a region of the radio frequency spectrum that is denied to Earth-based observatories because of the terrestrial ionosphere. Moreover, the antenna arrays required are so large in extent that it is not feasible to have them in orbit around Earth. The basin of the crater Tsiolkovsky (figure 7.13) has been suggested as a good location for such an array. The individual elements can be set down on the surface without preparation of the site, and this might be done by automated rovers because the elements are simple, small packages each consisting of a quadrupole antenna, a receiver, and a solar/battery power supply. The central collector of the system would be located on high ground, such as the central peak of Tsiolkovsy, from which communication would be maintained with the lunar outpost, possibly through a lunar orbital relay satellite, or repeaters on other lunar peaks.

Another important aspect of lunar basing for radio astronomy observations is to use a telescope on the Moon as one part of a long baseline interferometer with the other part being on Earth. This would provide a baseline of 248,000 miles (400,000 km). Such a configuration would provide a great increase in resolution over any array established solely on Earth or in close orbit around Earth.

As mentioned earlier, for a lunar radio-telescope array of many individual dishes, similar to the National Radio Astronomy Facility in New Mexico, a relatively flat site is required free of nearby mountains, rilles, or cliffs. Preferably this would be a lava filled crater on the far side of the Moon, perhaps just over the rim. A filled crater such as Wargentin would be ideal, but unfortunately this crater, although close to the limb of the Moon, is on the Earth-facing side, and a station there would have to suffer terrestrial radio interference.

FIGURE 7.12 The Moon also offers opportunities to set up arrays for detection of very low frequency radio waves, as shown in this sketch by a Jet Propulsion Laboratory artist, and also a system to detect the elusive gravity waves that scientists have been searching for. (*JPL*)

The Hubble Space Telescope in LEO provides optical telescope capabilities beyond those obtainable with Earth-based telescopes. A large telescope operating on the airless lunar surface, and shielded from the glare of Earth by choosing a location on the far side, can provide another quantum jump forward in optical astronomy. The Moon's surface location has several advantages over a location in Earth orbit. Night on the far side is the ultimate environment as far as the background of light is concerned. In LEO, however, the telescope has to contend with light from Earth, Sun, and Moon on a rapidly changing basis since the Hubble telescope moves completely around Earth in less than two hours, alternately in full sunlight and then in Earth's shadow. The observatory on the Moon would be outside the geocorona of the Earth so that the background of Lyman-alpha radiation would be reduced and longer exposures could be made at ultraviolet wavelengths.

An advantage of operation on the lunar surface is that a telescope can observe right down to the lunar horizon and the abrupt cutoff of light from the Sun at the horizon allows observation of the solar atmosphere and the corona, zodiacal light, comets, and the inner planets. As with the earlier comments about X-ray and gamma-ray astronomy, the predictable nature of the lunar motions, and their slow change, make it easier to guide the telescope system. Also, long uninterrupted exposures of up to 14 days can be given for very faint objects without any skylight fogging the images.

An important possibility is that of constructing a large optical telescope on the lunar surface in the form of multiple telescopes linked together to produce a telescope of gigantic proportions with enormous light gathering power and resolving power. Developed from the initial optical interferometer system mentioned earlier in connection with the JPL studies, such a telescope would rely upon sophisticated computer processing and integration of optical data from individual telescopes at a central control facility. It would have the ability to detect even terrestrial-sized planets around other star systems, observe markings on the surfaces of nearby stars, and inspect Solar System objects to see details equivalent or better

FIGURE 7.13 A suitable site for a radio astronomy observatory might be the crater Tsiolkovsky (a) on the lunar far side. This crater has a relatively large flat floor

than those so far seen with flyby spacecraft. The telescope, with its ability to probe optically to the virtual limits of the observable universe and back in time to its creation, has the potential to help solve the deepest philosophical problems that humans face.

In choosing a site for a lunar optical observatory it may be desirable to place it at an altitude where the haze from secondary ejecta would not be troublesome, since meteors hitting the Moon's surface stir up dust, some of which may enter orbit. A site close to, but a few degrees south of, the equator is preferable. This would permit observations of the Magellanic Clouds that lie close to the south pole of the heavens as seen from the Moon. The largest optical telescopes should be on the far side of the Moon, while smaller telescopes might initially be placed close to the limb regions so that they are periodically out of line-of-sight to Earth. Experience would be gained with these smaller telescopes before committing to the construc-

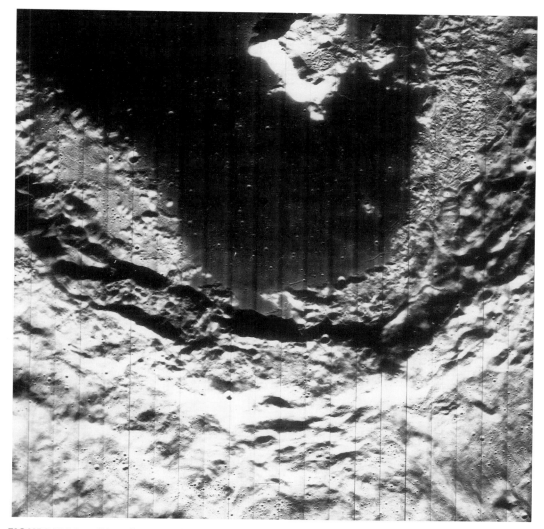

FIGURE 7.13 (b) and a central mountain group, which might be useful as a location for a laser relay tower to send data by laser links extending tower-to-tower back to the main base in the limb region without producing any radio interference from communications. (*NASA photos*)

tion of a large observatory on the far side since this would be a major architectural engineering task.

Another important use of the Moon for astrophysical research might be the establishment of a gravitational wave detector. Such a detector was studied by JPL scientists as one instrument that would benefit most from being located on the Moon. Gravitational waves travel at the speed of light from any massive object subjected to a varying acceleration. Despite a number of searches by various instrumental configurations here on Earth, such waves that are predicted from the general theory of relativity have not been recorded. The limitless vacuum, a quiet environment, and ability to use long laser paths on the Moon may enable scientists to detect the expected movement of suspended masses (detectors) resulting from gravity waves, movements that are predicted to be only a small fraction of the wavelength of visible light. Ultimately, a gravity wave detector on the Moon might be connected with another detector on Earth to record the gravity waves from exploding or collapsing stars.

• EXPLORATION: ITS OPPORTUNITIES AND CHALLENGES

We are a privileged generation to have experienced the exploration by pioneering unmanned spacecraft of nearly all the planets of the Solar System and most of their satellites and the landing of humans on another world. Living things are characterized by an ability to do everything possible to survive and to adapt to new environments. Human expansion into space is a new period of evolution for our species in particular, and perhaps for terrestrial life in general since we will undoubtedly take other living things with us to neighboring worlds of the Solar System. In addition a new viewpoint from space, looking out upon the cosmos from other worlds, will provide an unprecedented opportunity for us to answer fundamental questions about the nature of life itself.

On the Moon (figure 7.14) we will gain invaluable experience in operating closed ecological systems and efficient recycling systems not only for human expansion into the Solar System but also for feedback to terrestrial applications.

Gaining new information about our function as living beings and our capacity to adapt to different environments in space and on other worlds and to live and work away from Earth is vitally important in this age when natural or human-made catastrophes could affect the survival of our species. Despite years of futile attempts to achieve some control over nuclear armaments we are still threatened by an

FIGURE 7.14 On the Moon, humans will gain invaluable experience in living and working and operating many different types of machines on another world. (*NASA-Johnson*)

overkill of multiple warheads targeted at population centers as well as military installations. This situation does not appear to have any chance of changing in the near future despite talks between leaders of the two nations who have perpetrated this nuclear threat. While token reductions have been made, thousands of nuclear warheads remain as threats to all life on Earth, and nuclear testing by several nations continues to produce more effective nuclear weapon systems.

The assured survival of higher forms of terrestrial life may depend upon moving some of the genetic pool to other worlds and, following the establishment of outposts and permanent bases, creating settlements or colonies away from Earth. Perhaps way downstream we might experience a complete separation from terrestrial conditions of governance, an evolution similar to that which took place when the American colonists created a government of checks and balances as they rejected many traditional forms that had plagued Europe for centuries.

Biological questions of fundamental importance are currently being answered through spaceflight. How does life originate and evolve? How does life adapt to changing environments? How do living systems perceive and use gravity? How can humans operate effectively for long periods under conditions of reduced gravity or microgravity?

Seeking answers to such fundamental questions about human origin, evolution, and possible futures has always been important to the most advanced civilizations of our planet. The story of the evolution of the universe, of the Earth, and of humans has always fascinated humankind. It is, some thinkers have claimed, our moral and intellectual duty to research this story with every means at our disposal and to describe it in modern terms.

Our experience in working on the Moon will be fundamental to establishing a basis of understanding how to thrive on other worlds before expanding into the Solar System. This experience is necessary if we are to realize the potential economic wealth available from the Solar System's diverse planets, satellites, asteroids, and comets. But all benefits will not be measured materially. A new world viewpoint will expand creativity in the arts and a broadening of philosophy, a veritable second Renaissance.

Doubtless there are possibilities for underground lunar cities, specialized science centers, research centers, and manufacturing facilities with access to extensive high vacuum on the lunar surface, medical facilities and rehabilitation centers, and retirement homes in low gravity to aid the physically impaired. These are, at this stage, beyond the scope of this book's purpose. What will follow the establishment of bases is extremely speculative at present. While we can make many predictions about what might happen, the lesson of history is that we are usually hopelessly wrong in our predictions. There is always great danger in extrapolating into the future based on what is happening today and assuming that current conditions will continue unchanged. This error of thinking is often committed in the political or social scene, but it also applies to science and technology predictions. This writer remembers how he wrote some first technical papers on unmanned communications satellites in geosynchronous orbit ("Into Space," *Aeronautics*, November 1946; "The Establishment and Uses of Artificial Satellites," *Aeronautics*, September 1949), and assumed that they would use plastic bag solar mirrors to heat a working fluid for a closed cycle turbine generator of electrical energy from solar radiation (Figure

7.15). In fact, by the time big rockets were available to carry communications satellites into orbit, solar cells had been invented. They were readily available to make the power supply system much simpler than he had envisaged a decade earlier.

Over the years there have been several papers written on planetary engineering, sometimes called terraforming, in which the environment of an extraterrestrial body, such as Mars or Venus, might be changed to suit humans. Mars appears to be the most likely candidate for such activities and even so the replenishment of the Martian atmosphere and flooding of its ocean basins with water may take many hundreds if not tens of thousands of years. Venus appears to be an even more difficult planet for terraforming. As for the Moon, the low lunar gravity would allow most gases needed for a breathable atmosphere to escape quickly. Maintaining a breathable lunar atmosphere would be extremely difficult with any known technology. The most likely way to use planetary engineering on the Moon would be to construct large habitats beneath the surface or within covered dome structures sited in deep craters (figure 7.16) on the surface. The latter would always be exposed to hazards of incoming meteorites, some of which would be large enough to puncture the domes. Such ambitious activities are not the subject of this writing, but they cannot be discounted as impossible or never likely to be attempted. Humans seem to try all manner of difficult undertakings, often without any ob-

FIGURE 7.15 It is important to note that very often achieving a purpose works out to be much simpler than when originally conceived. The author proposed an unmanned satellite for telecommunications in 1949, as shown in this artist's concept. But actual communications satellites proved much simpler and did not need the closed cycle gas-turbine power supply he had proposed. The same may be true of the future missions to return to the Moon and establish a permanent human presence on our neighbor world. (*Painting by the author*)

FIGURE 7.16   Even further into the future, regular-shaped deep craters such as this one (Schmitt) on the western edge of Mare Tranquillitatis, might become sites for lunar surface cities, enclosed like astrodomes in protective covers. This crater is 7 miles (11.25 km) in diameter and could house a large community with its supporting agriculture. Inhabitants of such a lunar city would be free of any psychological problems that might arise from extended periods of living underground in lava tubes. (*NASA*)

vious economic advantage but purely to satisfy the desire to accept challenges and somehow overcome what at first seem unsurmountable difficulties.

Following the emergence of the human species from the terrestrial cradle, and its first hesitant crawling into cislunar space, the outpost on the Moon is a most important first upright step in the growth of the species toward adulthood. Let us not try to pull humanity back into its cradle. Instead, let us encourage the children of the space age, that half the human family born since the first humans walked on the Moon, to begin to walk confidently to a bright future, and to reach for the stars. For in the words of astronaut Alfred M. Worden, enthusing later about what it had been like when he piloted Apollo 17's Command/Service Module around the far side of the Moon, completely cut off from Earth: "The next step is out there. Out there stars shine, pieces of light . . . a pattern of so much brilliance that I am honored even here." He explained to this author, when we appeared together on a panel at a space enthusiasts' convention, that on the far side of the Moon, when in its shadow, the whole universe seemed ablaze with light.

The stars indeed are beckoning, and the problems of cosmology continue to tantalize the human mind. In the diamond-point brilliances of a night sky, when we are freed from the sense-obscuring light pollution of our cities, the stars appear to send us a silent challenge, a call to seek out and explore their mysteries and purpose.

We can, and we should, heed their call. A permanent human presence on the Moon will be our first affirmative answer.

GLOSSARY

ACCRETION: Collection of small bodies into larger bodies by mutual gravitational attraction.

AEROBRAKE: A shell-like structure used to protect a spacecraft and dissipate some of its kinetic energy so that it can enter LEO without having to use rocket propellants.

AGE DATING: Ascertaining the ages of rocks by analysis of proportions of different isotopes resulting from radioactive decay of various elements since the rock was formed.

AIRLOCK: Chamber to permit a spacesuited human to pass from a region of one atmospheric pressure to another; e.g., from the interior of a habitat to the vacuum of the lunar surface.

ANORTHOSITE: A plutonic rock consisting mainly of plagioclase.

ASTEROID: A small planetary body of which many thousands are known over a wide range of sizes; most orbit the Sun between the orbits of Mars and Jupiter and are a potential source of raw materials for human use.

ASTHENOSPHERE: Inner plastic region of a planetary body below the outer, rigid lithosphere.

ATRIUM CONFIGURATION: A configuration suggested in one NASA study for a manned satellite in LEO to assemble and service lunar and interplanetary spacecraft. Consists of several hanger-like rigid structures around a central open cubical area.

BALLISTICALLY EJECTED MATERIAL: Material ejected from a meteorite impact site at a sufficiently high velocity to follow a ballistic trajectory before landing back on the surface.

BASALT: A dark-colored, fine-grained volcanic rock.

BENEFICIATION: Processing of mineral ore to increase the percentage of mineral before the refining process.

BIOSPHERE: The shell surrounding a planet in which living things can thrive.

BRECCIA: A fragmented rock produced by the action of meteorite impacts.

CAMBRIAN ERA: A terrestrial period starting some 600 million years ago when from the fossil records the first marine invertebrates appeared.

CARBONACEOUS CHONDRITE: An asteroid or meteoroid composed mainly of carbon-rich materials.

CELLULAR ROCKET: A rocket made up of a number of individual rockets that when exhausted of propellant can be jettisoned to eliminate their dead weight and thus provide a higher mass ratio for the vehicle.

CELSS: controlled ecological life-support system; an advanced system to support human life on other worlds including production of food through agriculture, and recycling of everything possible.

CHONDRITE: A primitive, unmelted, and undifferentiated meteorite consisting of stony materials.

CISLUNAR: Space within the orbit of the Moon.

CONSTRUCTION SHACK: An initial installation on the lunar surface with integral power supply and life-support system from which workers can assemble and equip larger modules and bring them to operational status for a lunar outpost.

COPERNICAN PERIOD: The most recent period of lunar history extending from the time of formation of bright-rayed craters such as Copernicus to the present time.

COSMIC RAYS: Highly energetic subatomic particles originating from the Sun and from outside the Solar System.

CRYSTALLINE: Rocks consisting predominantly of mineral crystals; usually formed by precipitation from hot magma as it cools and solidifies.

DIFFERENTIATION: The gravitational separation of the material of a planetary body into shells with the heavier materials sinking toward the center.

DRIVE-THROUGH CONFIGURATION: A configuration for a large satellite in LEO to be used to assemble or service interplanetary or lunar spacecraft. As proposed in a NASA study it would be an open tunnel structure with spacecraft moving from an entrance through to an exit and being worked on during this passage through the station.

DUNITE: A rock that consists mainly of olivine.

DYNAMO EFFECT: Production of a magnetic field of a planet by a flow of electrical current within the planet.

ECLSS: Environment control and life-support system; a system to support humans with air and consumables in space stations, in spacecraft, or on another world.

ECONOMIC COMMONS: Derived from the old British concept of a common within a village where its resources belonged to all the people and no particular person. Resources that belong to the people in common and not to any government or commercial enterprise.

ERATOSTHENIAN PERIOD: The period of lunar history prior to the Copernican, when fresh craters were formed whose ray systems have since been obliterated by subsequent impacts.

EVA: Extravehicular activity; activity outside a spacecraft or a lunar habitat with an astronaut protected by a suit that provides a respirable atmosphere and protection from the ambient conditions.

FELDSPAR: A calcium-rich silicate rock.

FRACTIONATION: Fractional crystallization; formation at successively lower temperatures of various mineral components of a hot magma as it cools, so that some minerals because of their specific gravity tend to concentrate in a layer.

GABBRO: A plutonic rock; a coarse-grained basalt.

GAMMA RAYS: Highly penetrating radiation similar to X-rays; emitted by radioactive elements as they disintegrate.

GEOPHONE: An instrument to record vibrations transmitted through rocks.

GEOSYNCHRONOUS ORBIT: An equatorial orbit of sufficient radius for a satellite to revolve around the Earth in the same period Earth rotates on its axis, so it seems to hang above a given point on Earth's equator.

GRAVITATIONAL ANOMALIES: Variations from a regular gravitational field expected of a homogeneous or a symmetrical planetary body.

HABITAT: A structure placed on the Moon to house a permanent crew and provide them with living and working quarters including a control center for other modules and operations of a lunar outpost.

HYDROSPHERE: The shell of liquid water on a planet.

ILMENITE: A minor lunar mineral, possibly concentrated in some areas, that is a potential source for iron, titanium, and oxygen. Especially abundant in some types of mare basalts.

IMBRIUM PERIOD: The period in lunar history when the great mare basins were excavated.

IMPACT BASIN: A large circular depression surrounded by concentric mountain ranges caused by the impact of a body of relatively large size into a planet or satellite. Mare Orientale is an excellent example on the Moon.

INTERFEROMETRY: Use of the interference patterns produced when two beams of radiation overlap to produce alternating bands of addition and subtraction. Allows details to be resolved beyond the normal limits of a radio or optical telescope by analysis of the interference patterns produced.

IONOSPHERE: Region of ionized gases in the low density upper atmosphere of a planet resulting from incoming solar radiation producing ions and free electrons from the atmospheric gases.

ION PROPULSION: A propulsive device that relies upon an electrically accelerated stream of ions to provide an extremely high velocity propulsive jet.

ISOTOPE: Form of an element that although having the same chemical properties differs in the contents of its atomic nucleus so that the atomic mass differs because of the presence or absence of some neutrons.

KREEP: A lunar rock enriched in potassium (K), rare earth elements (REE), and phosphorous (P).

LAVA: Hot molten rock flowing or solidified from flows on the surface of a planet.

LAVA CHANNEL: A channel eroded by a fluid lava stream.

LEO: Low Earth orbit; an orbit just outside Earth's appreciable atmosphere.

LEV: Lunar Excursion Vehicle; a spacecraft designed for traveling between an orbit around the Moon and the lunar surface.

LIBRATION: Oscillations of the Moon relative to the observer arising from various causes that allows more than half of its surface to be seen over a period; includes longitudinal, latitudinal, and diurnal effects, the last named being produced by a difference in viewpoint when the Moon is seen rising or setting from Earth's surface.

LITHOSPHERE: The solid rocky outer layers of a planet below which is the magma of the interior.

LLO: Low lunar orbit; an orbit around the Moon at a convenient height above the surface for lunar landing and ascent vehicles to travel efficiently between the lunar surface and an orbiting spacecraft.

LTV: Lunar Transfer Vehicle; a spacecraft to transport crews or cargoes between LEO and lunar orbit.

LUNATION: Period from one New Moon to the next.

MAGMA: Hot molten rock inside a planet.

MAGNETIC PERMEABILITY: The ratio of the flux density produced to the magnetic force applied.

MAGNETOMETER: An instrument for measuring the strength of a magnetic field.

MAGNETOSPHERE: The region around a planet where the magnetic field of the planet is dominant and holds off the solar wind.

MAGNETOTAIL: The region of the magnetosphere extending from a planet like a wind-sock trailing down the solar wind.

MANTLE: The shell of a planetary body beneath the crust.

MARE: A dark-colored, relatively flat, lowland area of the Moon.

MASCON: A concentration of mass beneath some mare surfaces of the Moon.

MASS DRIVER: A propulsion system, first proposed by Gerard O'Neil, to use material from the Moon or an asteroid and accelerate it to form a propulsive jet.

METEOR: The glowing body or streak appearing in an atmosphere when a meteoroid is heated to incandescence by passage through that atmosphere. If the meteoroid is not completely consumed it arrives at the surface and is known as a meteorite.

METEORITE: A meteoroid that has fallen onto the surface of a planetary body.

METEOROID: A relatively small rocky or metallic body moving through space and thought to be the remnant of collisions of planetesimals, disintegrated comets, or residuals from the condensation of the primeval solar nebula.

MICROMETEOROID: An extremely small meteoroid.

MIR: A complex of vehicles and structures assembled by the Soviets in LEO to act as a permanent space station.

NORITE: A coarse-grained igneous rock consisting mainly of plagioclase and pyroxene.

OLIVINE: A dominant lunar mineral in mare basalts; a potential source of magnesium, iron, silicon, and oxygen.

PETROLOGY: The study of rocks; their origin, condition, composition, and effects of weathering processes

PHOTODYNAMIC SYSTEM: A system to generate electrical power from solar radiation which relies upon use of a working fluid acting through a turbogenerator.

PHOTOVOLTAIC SYSTEM: A system to generate electrical power from solar radiation which relies upon a solid-state device to produce a direct current output from banks of solar cells.

PLAGIOCLASE: A mineral group that forms a major part of anorthositic rocks, breccias, and a smaller part of mare basalts. It is a potential source of aluminum, silicon, silicate, and oxygen.

PLANETARY DYNAMO: See dynamo effect.

PLANETESIMAL: A body made up of condensates from a primordial nebula from which planets and their satellites formed by accretion.

PLATFORM CONFIGURATION: A suggested configuration for an orbital transportation node in LEO derived basically by adding a platform on a long vertical truss attached to a modified Freedom-type of space station. Interplanetary and lunar spacecraft would be assembled and serviced on the platform.

PLUTONIC ROCK: An igneous rock that crystallized at depth.

POLYMICT: Multiple rock types within a single sample; usually applied to breccias containing other lunar breccias.

PRE-BIOTIC MOLECULE: A carbon-rich molecule from which a living organism may have developed.

PYROXENE: A group of silicate minerals containing magnesium, iron, and with smaller amounts of aluminum, titanium, chromium, and sodium.

RADIATION BELT: Region of energetic particles trapped in the magnetic field of the planet.

RADIOACTIVE ELEMENT: An element that will naturally decay into simpler elements by release of particles from its atomic nucleus.

REGOLITH: The surface material of a planet, particularly one that has been mixed by the impact of bodies from space onto that surface.

REMANENT MAGNETISM: Magnetism remaining when the inducing magnetic field is no longer present.

RILLE: A fault trough or a sinuous channel

ROBOTICS: Activities involving machines and humans acting in concert; machines acting automatically to carry out preprogrammed tasks, or remotely under direct contol of a human being.

RTG: Radioisotope thermoelectric generator; a unit to provide electrical power from the conversion of alpha particle energy arising from radioactive decay of plutonium dioxide.

S-BAND TRANSPONDER: A radio device operating in a frequency band around 2.3 GHz used for determining the distance to the vehicle carrying the transponder.

SEISMIC WAVE: A vibration transmitted through rocks from a mechanical disturbance in those rocks such as an earthquake or an explosion.

SHIRT-SLEEVES ENVIRONMENT: An environment beyond Earth's atmosphere that enables humans to operate effectively without having to wear space suits.

SIDEREAL MONTH: The period in which the Moon returns to the same position relative to the stars as observed from Earth.

SIDEROPHILE: Elements that concentrate in metallic iron.

SILICATES: A wide group of minerals all members of which contain silica; rock-forming examples are feldspar and pyroxene.

SINTER: The development of bonds by heating to convert loose materials to solids.

SOLAR FLARE: A stormlike eruption on the Sun emitting intense ultraviolet and other radiation and energetic particles that are shot across the Solar System.

SOLAR STORM: A major disturbance on the Sun producing solar flares and eruptions of matter into space from the Sun.

SOLAR WIND: Stream of plasma (mostly protons and electrons) flowing outward from the Sun.

SPACE POWER SYSTEM: A system of generating electrical power from solar radiation at a power station in LEO, and transmitting the power to Earth.

SPACE SETTLEMENT: A large satellite in Earth orbit to house many humans on a permanent basis.

STACK: An assembly of modules to form a lunar spacecraft for transport of humans or cargo to or from the Moon.

STRATIGRAPHY: Basing the age of geological units by assuming that older units are overlain by younger units.

STREPTOCOCCUS MITIS: A microorganism common on Earth which survived for several years on the Moon inside the camera of a Surveyor spacecraft.

SUITPORT: A system developed for use in space station Freedom that allows a person to quickly don or exit from a spacesuit through a port in the back of the suit aligned with a port in the airlock chamber.

SYNODIC MONTH: Lunation; period from one New Moon to the next.

TECTONICS: Molding of a planetary surface by forces acting from within the planet.

TELEMETRY: Making measurements at a distant location and transmitting the data to another location for analysis.

TELEOPERATION: Operation at a distance; remote operation of a machine by commands transmitted to it from afar.

TELESURGERY: Surgery conducted from a distance by use of radio or other communication and teleoperators.

TRANSIENT EVENT: Events observed on the Moon lasting only for a short while and thought by some observers to represent outgassing from the lunar interior.

TRANSIENT MAGNETIC FIELD: A temporary magnetic field; in the case of the Moon it is caused by the solar wind.

TRANSPORTATION NODE: An artificial satellite stationed in LEO at which lunar or interplanetary spacecraft can be assembled, fueled, serviced, or retrofitted.

TROCTOLITE: A coarse-grained basic igneous rock consisting mainly of olivine and plagioclase.

V-2: German ballistic missile produced during the concluding phases of World War II, capable of carrying a one-ton warhead about 200 miles.

VESICULAR: A character of many extrusive igneous rocks in which there are many cavities some of which are lined with minerals. The vesicles (cavities) result from expansion of gases escaping from the rock as it cools.

WEATHERING: Changes to the surface of a planet by various processes such as wind erosion, water erosion, chemical actions, radiation, energetic particles, impact of meteorites.

X-RAY FLUORESCENCE: Effect of X-rays on materials giving rise to other X-rays of frequencies that can be used to determine the nature of the substance fluorescing.

BIBLIOGRAPHY

THE MOON, BACKGROUND

Baldwin, Ralph B. *The Face of the Moon.* Chicago: The University of Chicago Press, 1949.

Baldwin, Ralph B. *A Fundamental Survey of the Moon.* New York: McGraw-Hill, 1965.

British Astronomical Association. *Guide to Observing the Moon.* Hillside, N.J.: Enslow Publishers, 1986.

Gamow, George. *The Moon.* New York: Abelard-Schuman, 1953.

Hess, Wilmot N. Ed. *1967 Summer Study of Lunar Science and Exploration.* Washington, D.C.: NASA SP-157, 1967.

Markov, A. V., Ed. *The Moon, A Russian View.* Chicago: University of Chicago Press, 1962.

Proctor, Richard A. *The Moon.* London: Longmans, Green, 1898.

Singer, S. Fred, Ed. *Physics of the Moon,* Tarzana, American Astronautical Society, 1967.

Smith, R.A., and Arthur C. Clarke. *The Exploration of the Moon.* New York: Harper, 1954.

Various authors. *NASA 1965 Summer Conference on Lunar Exploration and Science.* Washington D.C.: NASA SP-88, 1965.

FICTION

Leighton, Peter. *Moon Travellers.* London: Oldbourne Book Co., 1960.

EXPLORATION BY SPACECRAFT

Bergaust, Erik, and Seabrook Hull. *Rocket to the Moon.* New York: Van Nostrand Company, Inc., 1958.

Burgess, Eric. *Assault on the Moon.* London: Hodder and Stoughton, 1966.

Caidin, Martin. *The Moon: New World for Men.* Indianapolis: The Bobbs-Merrill Company, Inc., 1963.

Collins, Michael. *Carrying the Fire.* New York: Farrar, Straus and Giroux, 1974.

Cortright, Edgar M., Ed. *Apollo Expeditions to the Moon.* Washington, D.C.: NASA SP-350, 1975.

Ertel, Ivan D., and Mary Louise Morse. *The Apollo Spacecraft, A Chronology.* Washington D.C.: NASA SP-4009 (3 vols.), 1969.

Gordon, Theodore J., and Julian Scheer. *First Into Outer Space.* New York: St. Martin's Press, 1959.

Green, Jack, Ed. *Interpretation of Lunar Probe Data.* Tarzana: American Astronautical Society, 1967.

Grimwood, James M., and Barton C. Hacker with Peter J. Vorzimmer. *Project Gemini, A Chronology.* Washington, D.C.: NASA SP- 4002, 1969.

Hall, R. Cargill. *Lunar Impact.* Washington, D.C.: NASA SP-4210, 1977.

Johnson, Nicholas L. Ed. *Handbook of Soviet Lunar and Planetary Exploration.* San Diego: Univelt, Inc., 1979.

Kosofsky, L. J. and Farouk El-Baz. *The Moon as Viewed by Lunar Orbiter.* Washington, D.C.: NASA SP-200, 1970.

Mailer, Norman. *Of A Fire on the Moon.* Boston: Little, Brown and Company, 1969.

Swenson, Loyd S., and James M. Grimwood and Charles C. Alexander. *This New Ocean, A History of Project Mercury.* Washington, D.C.: NASA SP-4201, 1966.

U. S. Congress, Office of Technology Assessment. *Exploring the Moon and Mars: Choices for the Nation.* Washington D.C.: OTA-ISC- 502, Government Printing Office, 1991.

U.S. News & World Report. *U.S. On the Moon: What it Means to Us.* New York: The Macmillan Company, 1969.

SPACE STATIONS

Belew, Leland F., and Ernst Stuhlinger. *Skylab, a Guidebook.* Huntsville: NASA Marshall Space Flight Center, EP-107, 1973.

Carter, L. J., Ed. *The Artificial Satellite.* London: British Interplanetary Society, 1951.

French, Bevan M. Ed. *Planetary Exploration Through Year 2000.* Washington, D.C., NASA Report (2 parts), 1986.

Froehlich, Walter. *Apollo-Soyuz.* Washington, D.C.: NASA EP-109, 1976.

Johnson, Richard D. and Charles Holbrow. *Space Settlements, A Design Study.* Washington D.C.: NASA SP-413, 1977.

Newkirk, Roland W., and Ivan D. Ertel with Courtney G. Brooks. *Skylab: A Chronology.* Washington, D.C.: NASA SP-4011, 1977.

Various authors. *Manned Space Stations* Symposium. New York: Institute of Aeronautical Sciences, 1960.

Various authors. *Apollo-Soyuz Test Project.* Washington, D.C.: NASA SP-412, 1977

APOLLO RESULTS

French, Bevan M. *The Moon Book.* New York: Penguin Books, 1977

Heiken, et al. *Lunar Sourcebook.* Cambridge: Cambridge University Press, 1991.

Taylor, Stuart Ross. *Lunar Science: A Post-Apollo View.* New York: Pergamon Press, Inc., 1975.

Various authors. *Biomedical Results of Apollo.* Washington, D.C.: NASA SP-368, 1975.

Williams, Richard J. and James J. Jadwick, Ed. *Handbook of Lunar Materials.* Washington, D.C.: NASA Ref. Pub. 1057.

Alred, John, Kriss J. Kennedy, Andrew Petro, Michael Roberts, Jonette Stecklein, and James Sturn. *Lunar Outpost*. Houston: NASA Johnson Space Center, Publ. JSC-23613, 1989.

Kent, Stan, Ed. *Remember the Future—The Apollo Legacy*. San Diego: Univelt, 1980.

National Commission on Space. *Pioneering the Space Frontier*. New York: Bantam Books, 1986.

O'Neil, Gerard K. *The High Frontier*. New York: William Morrow & Company, Inc., 1977.

Ride, Sally K. *Leadership and America's Future in Space,* Washington D.C. Internal Report to NASA Administrator, 1987.

Roth, Emanual M. *Bioenergetics of Space Suits for Lunar Exploration*. Washington, D.C.: NASA SP-84, 1966.

Ruzig, Neil P. *The Case for Going to the Moon*. New York: G. F. Putnam's Sons, 1965.

Space Science Board. *Lunar Exploration: Strategy for Research 1969–1975*. Washington, D.C.: National Academy of Sciences, 1969.

Stine, G. Harry. *The Third Industrial Revolution*. New York: G. P. Putnam's Sons, 1975.

Taylor, G. Jeffrey and Paul D. Spudis, Ed. *Geoscience and a Lunar Base*. Washington D.C. NASA Conf. Pub. 3070, 1990.

Toelle, Ronald. *Heavy Lift Launch Vehicles for 1995 and Beyond*. Huntsville: NASA Marshall Space Flight Center, Tech. Memo. 86520, 1986.

Various authors. *Post-Apollo Lunar Science*. Houston: Lunar Science Institute, 1972.

Various authors. *Technology Needs for Human Exploration Missions*. Washington, D.C.: NASA Internal Report by Office of Exploration, 1989.

Various authors. *Proceedings of the Sixth Annual Meeting of the Working Group on Extraterrestrial Resources*. Washington, D.C.: NASA SP-177, 1968.

Various authors. *Report of the 90-day Study on Human Exploration of the Moon and Mars*. Washington D.C.: Internal Report to NASA Administrator, 1989.

Williams, Louis J., Terrill W. Putnam and Robert Morris. *Aeroassist: Key to Returning from Space and the Case for AFE*. Washington, D.C., NASA Publication, 1991.

Williams, Richard J., and Norman Hubbard. *Report of Workshop on Methodology for Evaluating Potential Lunar Resource Sites*. Houston: NASA Johnson Space Center, Tech. Memo. 58235, 1981.

INDEX

Olivine, 100, 102, 112, 230, 254

Olympus, xii

Olympus Mons, 161

Omni Systems, Inc., 136

Operations, beyond Earth, 21; transportation node, 191

Opportunities, 246

Optical, interferometer, 241; telescopes, 239, 240, 243

Orbital, mapping, 52, 140; measurements, 104, 105; nodes, 171; sensors, 79, 101; speed, 27; transportation node, 184

Orbital Module, 179

Orbital Transfer Vehicle, 193

Orbital Workshop (see Skylab)

Orbiters, robotic, 131

Orbiting solar mirrors, 232

Orbits, 28

Orientale basin (see Mare Orientale)

Origin, anomalies, 105; craters, 10, 37, 93; magnetic fields, 112; mare basins, 37, 93; Moon, 17, 35, 36, 94, 121, 122

Orthopyroxene, 112

Outpost, 129; construction, 152, 198; development, 170; establishment, 170; plan, 200; site selection, 199

Outpost II, 175

Outputs, human, 235

Oxygen, 94, 95, 102, 146, 147, 148, 155, 159, 186, 192, 200, 220, 221, 223, 226, 228, 230, 234; plant, 221

Pacific Ocean, 35

Pacing item, HLLV, 186

Palus Putredinis, 44, 45, 76, 99, 100

Pantheons, xii

Parachute recovery, 66

Paramedic robots, 218

Paris Lunar Atlas, 31

Paris Observatory, 31

Parry, 73

Parsons, J. W., 50

Passage to India, 27

Passive seismic experiments, 106

Pathway to human outpost, 144

Pattern of new ideas, xi

Payload required, 184

Peenemunde, 82

Penetrator probes, 138, 139, 152

Pennsylvania State University, 148

Perigee, 23, 28

Perihelion, 28

Period, 27; of cratering, 37; of exposure, 118, 119, 120; of volcanism, 110

Personnel shelter, 158; transportation system, 171

Petrology, 254

Phases, 23, 27, 28, 117

Philosophical questions, 17

Philosophical Magazine, 3

Phosphorous, 95

Photo mapping, 52, 131

Photodynamic system, 207, 208, 254

Photography, 31, 65, 128

Photovoltaic system, 207, 208, 254

Physical explorer types, 219

Pic du Midi Observatory, 31, 162

Pickering, W. H., 31

Pinewood Studios, 27

Pioneer, 6, 8, 49

Placement of habitat, 206

Plagioclase, 95, 100, 109, 230, 254

Plagioclase feldspar, 95, 102

Plains, 23

Plan, lunar outpost, 201

Planetary dynamo, 254

Planet, ix; types, 27

Planetary, engineering, 248; evolution, 9, 26, 145, 146; magnetic fields, 113; quarantine, 216; resources, 146; systems, 243

Planetesimals, 36, 110, 255

Plant growth unit, 234

Plants, 234

Plasma, 87

Plasticity, asthenosphere, 107

Plates, formation, 17

Platform configuration, 190, 191, 192, 255

Platinum, 94

Plato, 43

Plutarch, 24

Plutonic rock, 95, 97, 255

Plutonium dioxide, 69

Podgorny, N., 71

Poe, E. A., 24

Polar, ice, 161, 164; orbiting mission, 139; volatiles, 159, 161, 164

Polar Orbiter, 161

Political leadership, 57; stability, 229

Polymict breccia, 99, 255

Porosity of surface, 47

Post-Apollo missions, 132, 133

Potassium, 87, 94, 95, 115

Potocnik, Captain, 172

Power, supply, 159; system, 231; habitat, 207; nuclear, 209, 210; solar, 207, 209

Pre-biotic molecule, 255

Precambrian era, 12

Precursor mission, 131, 164

Predictions, xi

Predictive unreliability, 247

Prefabricated modules, 156, 197, 198

Pressure, lunar atmosphere, 115

Pressurized busses, 200

Prinz, 225

Priorities, 145

Private enterprise, xii, 135, 136, 228; race to Moon, 137

Privileged generation, 246

Probes, 138, 139, 152

Probing into Moon, 106

Problems, structural, 239

Process of mining, 223

Processing, 147, food, 234; lunar rocks, 230; station, 144, 145; waste materials, 234

Proclus, 161

Project Horizon, 156

Promise of space expansion, xiii

Propellant, 186, 191, 192; consumption, 61; production, 147, 148; storage, 191, 192; tanker, 196

Propulsion, 148; module, 195; systems, 220

Prospector, 135

Protecting lunar samples, 15

Protection, 198; astronauts, 216; from dust, 205, 208;